SpringerBriefs in Geography

More information about this series at http://www.springer.com/series/10050

Gabriel Weiss · Erik Weiss
Roland Weiss · Slavomír Labant
Karol Bartoš

Survey Control Points

Compatibility and Verification

Springer

Gabriel Weiss
Institute of Geodesy, Cartography and GIS
The Technical University of Košice
Košice
Slovakia

Erik Weiss
Institute of Earth Resources
The Technical University of Košice
Košice
Slovakia

Roland Weiss
Institute of Earth Resources
The Technical University of Košice
Košice
Slovakia

Slavomír Labant
Institute of Geodesy, Cartography and GIS
The Technical University of Košice
Košice
Slovakia

Karol Bartoš
Institute of Geodesy, Cartography and GIS
The Technical University of Košice
Košice
Slovakia

ISSN 2211-4165   ISSN 2211-4173   (electronic)
SpringerBriefs in Geography
ISBN 978-3-319-28456-9   ISBN 978-3-319-28457-6   (eBook)
DOI 10.1007/978-3-319-28457-6

Library of Congress Control Number: 2015958902

Printed on acid-free paper

This Springer imprint is published by SpringerNature
The registered company is Springer International Publishing AG Switzerland

Preface

Geodesy, in its practice and in solving problems, should always take into account the spatial variability of its points in time, which are directly or indirectly related to technical activities and geodynamics of the Earth. Even the reference data related to survey marks of survey control points as "invariable values" may change over time (additional modifications, corrections in geodetic bases and map projections, etc.). Since it is necessary that points and data related to them should always be in a geometric consistency (i.e., points are compatible) for the reliable functionality of geodetic controls, it is also necessary to verify, or rectify, this conformity according to the user needs and circumstances.

The present work is devoted to these issues for geodetic controls on a local scale. Issues of compatibility of points, whose coordinates are expressed as functions of time by specific equations with respect to basal positions of points at certain epochs (points in systems ITRS, ETRS, and others), are not considered and solved in given issues.

The work is dedicated to all professionals who need to examine practically the stability of fixed survey control points in their various local formations as well as to students of Geodesy fields of study. The topic of verification of condition of survey control points is currently, when geodetic controls are and will be realized by various technologies and from different data, very timely. We hope that the present work will be useful for surveyors and provide them a sufficient overview of this issue.

Gabriel Weiss
Erik Weiss
Roland Weiss
Slavomír Labant
Karol Bartoš

Contents

Abbreviations

Bpv	Baltic Vertical Datum—after adjustment
S-JTSK	Datum of Uniform Trigonometric Cadastral Network
S(LOC)	Local Coordinate System
ŠPS	National Spatial Network
ŠTS	National Triangulation Network

Chapter 1
Introduction

The compatibility (homogeneity) of survey control points in established geodetic controls, and therefore the compatibility of relevant planimetric and altimetric networks (national, regional, local), is an important quality feature of them, which along with other quality parameters (accuracy, reliability, etc.) determine the usability of network structure for all challenging geodetic activities. What a compatibility of survey control points actually is and what is the scope of this concept is specified in Chap. 2.

The compatibility of each geodetic network depends not only on technology, methodology, and quality of its establishment, i.e., the accuracy of determination of its points in the relevant coordinate (vertical) system, but also on temporal changes of its physical points to a significant extent, which inevitably occur for various reasons depending on geodynamic or local geotechnical stability of the area, in which points are established. The compatibility, or incompatibility, of the existing geodetic control is also significantly created by the quality of determination of coordinates of physical survey marks of points.

Every geodetic control is characterized by a certain degree of incompatibility that gradually changes over time (displacements of physical survey marks of points from the initially surveyed positions arise, erroneous determination or adjustment of coordinates of some points occurs, datum inhomogeneity in the determination of new points increases, etc.) in the expansion, completion, and densification of structure of the original geodetic control since the beginning of its establishment, thus already in its initial structure. Also, the incompatibility of connecting and determined points resulting over time will always act in every completed, expanded part of geodetic control in varying degrees.

Therefore, it is absolutely necessary that the condition of compatibility of points to be used currently (as connecting, datum, setting-out, and other points) is thoroughly examined in the establishment of new geodetic control, respectively, when using points of already established networks. This means to identify incompatible, i.e., currently unusable points, which would deteriorate the quality of currently established points in new network formations when used.

This publication deals with the outlined issues of geodetic network structures, i.e., verification methods of the condition of existing geodetic controls regarding their compatibility and thus their usability. This geodetic issue is actual not only in

© The Author(s) 2016
G. Weiss et al., *Survey Control Points*,
SpringerBriefs in Geography, DOI 10.1007/978-3-319-28457-6_1

Slovakia, especially in relation to the quality of structure and homogeneity of ŠTS in the system of S-JTSK (Böhm et al. 1981; Michalčák et al. 1978; Vykutil 1982; Cimbálnik 1978 and others), but practically in every country (Hofmann-Wellenhof 1997; Hanke 1988; Lachapelle et al. 1982; Erker 1997 and others). The need for verification of compatibility is current not only when using terrestrial methods of point determination but also when using other surveying technologies since all technologies operate to a certain threshold accuracy and with physical survey marks of survey control points, located on an unstable earth surface.

The issue of compatibility is closely related to the establishment and processing of geodetic networks, while knowledge from this field is not presented in this work; we assume that they are well known to the reader.

Chapter 2
The Compatibility of Geodetic Control

Every survey control point (horizontal and vertical) has a physical mark and assigned numerical data (coordinates and heights) from certain reference systems, which should be related to a survey mark of this point. So far as those determining attributes are in stochastic planimetric or altimetric identity, the point can be considered as compatible, otherwise if specified attributes of point are mutually in geometrical significantly different positions, it is an incompatible point.

The compatibility of points can, therefore, be described as follows:

Compatibility is a characteristic of survey control point expressing the geometric stochastic relationship, i.e., a practical geometric identity of **physical mark** and a point defined by **corresponding data** of reference systems.

The mutual consistency or inconsistency of both characteristics of point might not significantly change over time, but it may often change due to various reasons (Sects. 3.1, 3.2) to such an extent that the point becomes practically unusable. Physical marks of points are fixed to the earth's surface that is continuously in motion, and numerical data of point, which should be permanent, may also change by new measurements and processing of geodetic control, adjustment of coordinate systems, datums, etc.

The phenomenon of compatibility of points or geodetic controls is sometimes also referred to as the homogeneity of points or geodetic controls. However, such a designation belongs to points rather in terms of the genesis of their coordinates (Böhm et al. 1981; Vykutil 1982). In this context, homogeneous (uniform) points are points whose coordinates in particular space resulted from the common measurement with a homogeneous accuracy, in a single type of processing (adjustment), with a single datum, etc. Non-homogenous (non-uniform) points in particular space are points that resulted from mixing up different types of homogenous points, with coordinates changed by various modifications of "neighborly relations" between partial homogeneous fields, while these conditions may be superimposed even by different changes–modifications in a cartographic projection, datums, and coordinate systems.

Characteristics of compatible and problems of incompatible points and geodetic controls are related to all types of standard geodetic networks with geometrical parameters, i.e., 1D altimetric networks (level and trigonometric), 2D planimetric networks, as well as 3D spatial networks.

© The Author(s) 2016
G. Weiss et al., *Survey Control Points*,
SpringerBriefs in Geography, DOI 10.1007/978-3-319-28457-6_2

The largest problems, in terms of compatibility of points in the current geodetic controls, arise and are presented in 2D planimetric networks, significantly less in 1D altimetric, mainly in level networks.

The incompatibility of points is most apparent in 2D geodetic controls. In this regard, the quality of geodetic controls is the most problematic for planimetric networks in all countries. In addition to the already outlined standard causes of incompatibility of points, also the specific local influences support its creation.

The issue of compatibility of height points in level networks occurs only occasionally. In terms of their genesis, it occurs practically only because of the change of physical marks of leveling points forced by the movements of their monument.

The issue of compatibility of points does not yet practically occur in currently constructed 3D networks although the current and future trends of the establishment of 3D and multidimensional geodetic controls will certainly require a creation of methods also for identification of 3D incompatible points.

Therefore, in this treatise on the incompatibility of survey control points, the most attention will be given to 2D geodetic controls, and the issues of compatibility in leveling point structures will be pointed out only to a lesser extent.

Chapter 3
The Compatibility of 2D (Planimetric) Points

3.1 Characteristics of Planimetric Points in Terms of Compatibility

Consider the position of points defined by coordinates $C = [X, Y]$ in a specific planimetric system $S(XY)$, which are properly monumented on the earth surface or objects. It is obvious that each point, as stated in Chap. 2, must be identified and defined by the following two fundamental components in terms of its functionality:

- physical position (monument with a survey mark) that is bound to the earth surface or objects attached to the surface and is variable in time (a point continuously changes its position in relation to close or distant surroundings with different speed as a result of various forces acting on lithospheric plates and in surface rock formations due to construction impacts, etc.);
- coordinates $C = [X, Y]$ of physical mark of point in a certain planimetric system $S(XY)$ that allows computing as well as various geodetic activities using the corresponding point. This component of point might also not prove to be stable over time; it varies mainly according to changes in the official geodetic control of the country concerned induced by the additional corrections (Fig. 3.1).

The correlation between these two components of the point should meet the basic requirement for the quality of point in term of compatibility, so the physical and coordinate positions of point correspond to each other at any time, i.e., so that both components of point have a stochastic positional identity in the used coordinate system. Simply expressed, coordinates are related accurately to the physical mark of point, respectively, and the mark in terrain corresponds to relevant coordinates from the $S(XY)$ system.

The compatibility of a planimetric point can also be formally characterized by point congruency of determinants (components) of the point. The point will be congruent, i.e., compatible, if the physical mark of point will always stochastically coincide with the geometrical point, formed by coordinates $C = [X, Y]$ in the ρ_{XY}

© The Author(s) 2016
G. Weiss et al., *Survey Control Points*,
SpringerBriefs in Geography, DOI 10.1007/978-3-319-28457-6_3

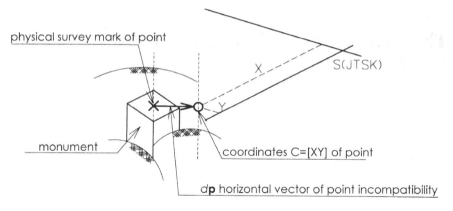

Fig. 3.1 Planimetric survey point and its two components (survey mark, coordinates)

plane of the $S(XY)$ system, at the time. For a different, i.e., mutually non-coinciding situation of both components of the point, when a statistically significant horizontal vector of incompatibility is formed between them (with a centimeter or even greater length), the point is identified as an incompatible point of geodetic control.

A geodetic control, i.e., a set of points in particular space that contains only compatible points is referred to as a *compatible geodetic control*. If a geodetic control also contains incompatible points, it is identified as an *incompatible geodetic control*, namely according to the quantity and distribution of incompatible points in the network (if known) as a moderately, strongly, locally, etc. incompatible geodetic control.

3.2   The Emergence of Incompatibility of Points and Its Impact on the Creation of Geodetic Controls

In a certain geodetic control, if incompatible points are presented or demonstrated, such geodetic control should be considered as a geodetic control of insufficient quality, in which incompatible points have to be identified in order not to be used for geodetic activities in that area.

The emergence of incompatibility of points in the existing geodetic controls established in the past by the system of stage construction is, however, an inevitable natural phenomenon that was generally demonstrated already in the establishment of initial points and changed with the continuous completion of geodetic control and with the trend of local expansion of incompatible points.

The verification of compatibility of geodetic control represents the verification of stability over time, and incorrect determination (coordinate) of its points and the character of geodetic control is always related to a certain time epoch in terms of compatibility, as is apparent from the genesis and development of incompatibility defects.

The emergence of incompatibility of survey control points in the establishment of geodetic controls can be schematically illustrated and explained by using Fig. 3.2, where time (epoch) expansion of the initial geodetic control (the initial network structure) from the original epoch t_0 is illustrated.

In the area of t_0 epoch, the scheme illustrates that all points have small vectors of incompatibility. Therefore, the linear non-identity of the physical and coordinate position of points has only a stochastic character.

In the t_1 epoch, determined points indicate that the vector of incompatibility at the point B_1 has a non-stochastic, significant size (e.g., it originated from a poor quality, inaccurate measurement between physical positions of points A and point B_1), while the vector of incompatibility at the point B_2 has a stochastic character.

In the t_2 area, the scheme illustrates that due to progressive deterioration of compatibility of points A and B, to which points C are connected, these points show significant incompatibility. There is also illustrated that physical positions of points C_2 and C_3 at the inter-epoch periods $t_2 - t_3$ shifted due to certain causes (e.g., anthropogenic, geodynamic, and various endogenous activities in that area).

The next t_3 epoch of geodetic control expansion shows that the trend of emergence of incompatible points is increasing, although for progressive connection of

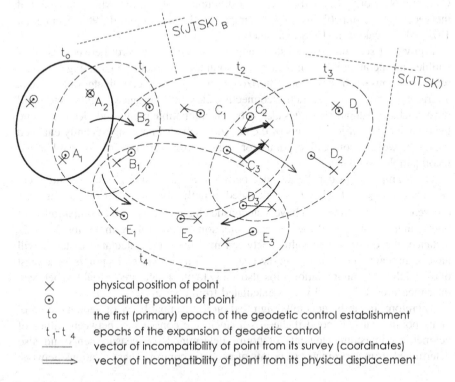

✕	physical position of point
⊙	coordinate position of point
t_0	the first (primary) epoch of the geodetic control establishment
t_1-t_4	epochs of the expansion of geodetic control
———	vector of incompatibility of point from its survey (coordinates)
⟹	vector of incompatibility of point from its physical displacement

Fig. 3.2 A geodetic control established at epochs (t_0, …, t_4)

points to the so far established network, for the part of new points, also a suitable compatibility can be formed for them due to a random, mutual elimination of various negative positional and surveying impacts.

This situation is also outlined for the t_4 area, in which the E points were determined from points B and D (epochs t_1 and t_3), respectively.

The incompatibility of determined points in a new epoch is also caused by the use of different datums for their determination. This situation is illustrated in the Fig. 3.2 in the t_4 epoch, in which the E points were determined from the t_1 and t_3 epochs and datum points B and D also determined from different epochs (t_0, t_2, i.e., from points A and C), define various implementations $S(XY)_B, S(XY)_D$ of coordinate frame of the system. As a result of point E determination with these different datums, errors together with the influence of measurement uncertainty also create their incompatibility will occur in their coordinates.

Points with suitable and unsuitable compatibility are the result of such creation of geodetic control, which is the most general procedure of its establishment.

In general, the incompatibility of points of certain geodetic control or its parts has a variable character, and it occurs in the space of geodetic control, mostly in certain regions, to different extents and with varying degrees of incompatibility. For example, also the incompatibility of points in the National Spatial Network—ŠPS of the Slovak Republic with coordinates of the Datum of Uniform Trigonometric Cadastral Network—(S-JTSK) varies for different local regions, while the length of the vector of incompatibility (3.7) averages 15–25 mm (Vykutil 1982; Cimbálnik 1978; Michalčák et al. 1978 and others).

Equivalent semantic relationships apply in a geodetic control between its compatible and incompatible points and corresponding geometric connecting elements (distances, observed directions, horizontal angles, and coordinate differences). Therefore, the relevant geometric connecting elements (distances, angles, etc.) will have stochastic, unbiased values also between compatible points and vice versa, and the relevant geometric elements of the network will also have significantly changed values between incompatible points (or even in configurations of compatible and incompatible points).

For example, consider the distance between points A_1 and E_1, namely the measured value $d_{A_1E_1}$ and value $d'_{A_1E_1}$ calculated from the corresponding coordinates, the difference between them will have a stochastic character (due to the compatibility of both points). However, if the converse situation occurs, for example, for the B_1E_2 distance, the difference of values between measured and calculated distance will have significant value, indicating the incompatibility of B_1 and E_2 points, or at least of one of them. Similar relationships also apply to other measured variables, between their measured values and values calculated from coordinates.

Therefore, it can be generally signified on the compatibility of geodetic control or its points that it is good if there is a stochastic consistency between values of geometric connecting elements of points determined by measurements and also determined from coordinates of points of geodetic control. These relations between elements of 2D networks may be well utilized in specific assessment and analysis of the compatibility of given geodetic control.

Consequently, the above analysis also indicates that the incompatibility of points is basically caused by:

- above the threshold *uncertainty* of determination (erroneous determination) of coordinates of the point's physical mark that results not only from the measurement itself and the influence of other various causes (determination of points of geodetic control using different datums, the use of heterogeneous surveying technologies, configuration defects in a geodetic control, etc.) but also from the processing of measurements;
- movement of the point's physical mark caused by various forces acting on the point in a given area.

Therefore, it is reasonable and realistic, when using any geodetic control and its local parts established in the past, to treat them as potentially incompatible geodetic control (with varying degrees of incompatibility of points and with their different dislocations and spread in a geodetic control), whose specific condition should be evaluated by appropriate procedures, when dealing with current surveying tasks.

If one uses a geodetic control or its part without any verification of the compatibility of relevant points, mainly the following negative consequences can be expected:

- Geometric connecting variables of points determined both from measurements and coordinate calculations may have significantly different values; however, one can identify areas with incompatible points by their localization.
- Although the adjusted network structure of new determined points will be consistent, but only with respect to the coordinates of used datum points, thus the adjusted coordinates of each determined point may not correspond to its physical position in an acceptable level.
- In the determination of new points using datum points from different areas of the network (with their expected incompatibility), non-identical coordinates of determined points that may not be at the same time compatible even with physical positions of determined points are obtained.

3.3 The Need for Verification of Compatibility of Survey Control Points

As mentioned above, the correlation between physical and coordinate positions of each survey control point in a network is characterized by a 2D vector of incompatibility $d\mathbf{p}$ ($|d\mathbf{p}|$ = a horizontal distance between the physical and coordinate positions of point in the ρ_{XY} plane) of unknown size and orientation; hence, the $d\mathbf{p}$ vector has a certain unknown, nonzero value (Sect. 3.4.2). In terms of compatibility of points, the size of the vector is assessed according to whether it is within designated or practiced accuracy of point determination or it deviates from these limits. According to that, the real, always existing incompatibility of each point can be assessed either as:

– a stochastic incompatibility, which characterizes the natural, practically unchanged condition of point (between the physical mark and its coordinates), only such points of geodetic networks should be used to deal with geodetic problems,

or as:

– considerable (significant) incompatibility, which characterizes the unnatural condition of point, when the vector of incompatibility $d\mathbf{p}$ exceeds the limit accuracy that is valid and used to determine new points. It is evident that points with significant incompatibility, i.e., incompatible points, of geodetic control should not be used.

Naturally, in terms of creating new groups of points with their connection to points of an old geodetic control (within an enlargement, densification, etc., of the local or superior network), only compatible points are suitable for these problems.

3.4   Indicators of Planimetric Compatibility of Points

3.4.1   In General

As follows from the definition and characterization of compatibility of points in the previous section, the compatibility of specific, single point in a geodetic control at the epoch t and without the results of its survey in another epoch t' cannot be objectively examined by geodetic methods. Therefore, it is necessary to have the information and data to examine the compatibility in a certain geodetic control:

– either from measurements and processing at different epochs t and t' ($t' > t$),
– or from an appropriate documentation of the epoch t and measurement and processing only at the epoch t'.

The compatibility of points is always determined for the time t' concerning their condition at the time t. Then, based on the comparison of:

– values of geometric elements of the network structure of geodetic control (distances, horizontal angles, etc.) between points from epochs t and t', i.e., determination of their differential–observational indicators

$$dL = L - L'(L = \{d, \omega, \ldots\}, L' = \{d', \omega', \ldots\}), \tag{3.1}$$

or:

– values of determined coordinates of points at epochs t and t', i.e., determination of their differential-coordinate indicators

$$dC = C - C', (C = f(L, \ldots), C' = f(L', \ldots)), \tag{3.2}$$

one can determine a compatible point (points), whose physical and coordinate positions have not changed over time $t' - t$, or incompatible point, whose physical position has moved or was determined by erroneous coordinates (Fig. 3.3) in some epoch (most often at t), using additional logical and mathematical analysis.

Therefore, if one can determine values of observational indicators dL or coordinate indicators dC of points of geodetic control or geometric relationships between points, these differences can be considered as indicators of compatibility, respectively, incompatibility, of point (points), on the basis of information contained therein. In a simplified view, low values in dL and dC will demonstrate their

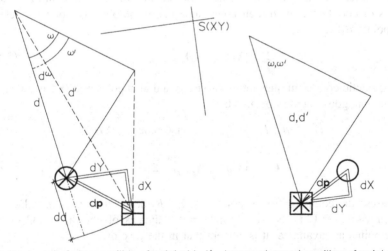

Displacement of phys. position of point at t - t´   Incorrect coord. position of point at t

✕   physical position of point at the epoch t

◯   coordinate position of point at t

⊥   physical position of point at the epoch t´   ——— measurement at t

▢   coordinate position of point at t´   - - - - measurement at t´

Fig. 3.3 Graphical representation of dC (dX, dY) and dL (dd, $d\omega$) in both types of incompatibility

compatibility, higher values their incompatibility. Naturally, the assessment of "lowness" or "highness" of values in the vectors $d\mathbf{L}$ and $d\mathbf{C}$ must be realized objectively, by appropriate mathematical procedures, based on the probability theory, mathematical statistics, and decision theory (Hald 1972; Anděl 1972, Riečan et al. 1983; Radouch 1983; Skořepa and Kubín 2001 and others).

3.4.2 Types of Indicators

In a given geodetic control, indicators $d\mathbf{L}$ and $d\mathbf{C}$ are formed from measurements performed in the network of geodetic control in two separate epochs t and t', where t represents the time of primary (initial) measurement of the geodetic control and t' represents the time, at which one want to have information on the condition of compatibility of geodetic control.

Let geometrical variables with values and their covariance matrix be measured at the epoch t in the geodetic control:

$$\mathbf{L} = [d_1, d_2, \ldots \omega_1, \omega_2, \ldots \text{ and others}]^{\mathrm{T}}, \mathbf{\Sigma}_{\mathbf{L}}. \tag{3.3}$$

Moreover, after the adjustment of the network structure of geodetic control, let the estimates of coordinates with their covariance matrix be obtained for its p determined points $B_1, \ldots, B_p$:

$$\hat{\mathbf{C}} = \left[\hat{X}_1 \hat{Y}_1 \ldots \hat{X}_p \hat{Y}_p \right]^{\mathrm{T}}, \mathbf{\Sigma}_{\hat{\mathbf{C}}}. \tag{3.4}$$

The geodetic control in question is surveyed and also processed at the epoch t', in which analogous data are obtained:

$$\mathbf{L}' = \left[d_1', d_2', \ldots \omega_1', \omega_2', \ldots \text{and others} \right]^{\mathrm{T}}, \mathbf{\Sigma}_{\mathbf{L}'} \tag{3.5}$$

$$\hat{\mathbf{C}}' = \left[\hat{X}_1' \hat{Y}_1' \ldots \hat{X}_p' \hat{Y}_p' \right]^{\mathrm{T}}, \mathbf{\Sigma}_{\hat{\mathbf{C}}'}. \tag{3.6}$$

To assess the compatibility of points $B_1, \ldots, B_p$ at the time t', both values of measured variables $\mathbf{L}, \mathbf{L}'$ between points and estimates of coordinates $\mathbf{C}, \mathbf{C}'$ of relevant points are available. It is evident that in the case of:

- physical stability of point's marks in the time period $t' - t$ and
- determination of points at the epoch t and t' with standard accuracy (without measured variables with "deviating values"),

differences $d\mathbf{L} = \mathbf{L}' - \mathbf{L}$ as well as differences $d\mathbf{C} = \mathbf{C}' - \mathbf{C}$ will represent small stochastic values, corresponding to the accuracy of measurement and adjustment. In that case, the geodetic control and all of its points can be considered as compatible, while there were no significant defects in a physical or coordinate component of points in this geodetic control during the periods $t' - t$.

In the case where the physical mark on the point B_i of geodetic control was shifted over time $t' - t$ or its coordinate position was erroneously determined at the time t (most commonly) or eventually at the time t', the corresponding differences of measured variables $dL_i, \ldots$ and coordinates $dC_i, \ldots$ of epochs t and t' will form large, more than expected, non-stochastic values; i.e., they will have the character of statistically significant values. Subsequently, based on the statistical confirmation of the character of corresponding differences $dL_i, \ldots$ and $dC_i, \ldots$ the relevant point B_i will be considered as an incompatible point of geodetic control for the epoch t' and remaining points, with stochastic differences $d\mathbf{L}$ and $d\mathbf{C}$, as compatible points.

Therefore, based on the size of values of observational $d\mathbf{L}$ and coordinate $d\mathbf{C}$ differences, it is possible:

$$
d\mathbf{L} = \begin{bmatrix} d'_{ij} \\ \vdots \\ \omega'_{kij} \\ \vdots \end{bmatrix} - \begin{bmatrix} d_{ij} \\ \vdots \\ \omega_{kij} \\ \vdots \end{bmatrix} = \begin{bmatrix} dd_{ij} \\ \vdots \\ d\omega_{kij} \\ \vdots \end{bmatrix}, \tag{3.7}
$$

$$
d\mathbf{C} = \begin{bmatrix} \vdots \\ X'_i \\ Y'_i \\ \vdots \end{bmatrix} - \begin{bmatrix} \vdots \\ X_i \\ Y_i \\ \vdots \end{bmatrix} = \begin{bmatrix} \vdots \\ dX_i \\ dY_i \\ \vdots \end{bmatrix}, \tag{3.8}
$$

or relevant planimetric differences—horizontal vectors of incompatibility $d\mathbf{p}$ at points:

$$
d\mathbf{p} = \begin{bmatrix} \vdots \\ dp_i = \sqrt{dX_i^2 + dY_i^2} \\ \vdots \end{bmatrix}, \tag{3.9}
$$

that directly characterize planimetric (coordinate) relations between epochs t and t', assuming its compatibility or incompatibility for the epoch t'. Differences $d\mathbf{L}, d\mathbf{C}(d\mathbf{p})$ that are quantitatively observable variables can, therefore, be considered as apposite, suitable indicators of compatibility of points.

3.4.3 The Internal Structure of Indicators

To use $d\mathbf{L}$ and $d\mathbf{C}$ as indicators for the assessment of compatibility of points, it is useful to know their structure, the character of components from which their numerical values are formed, as well as their properties, so that it can be concluded

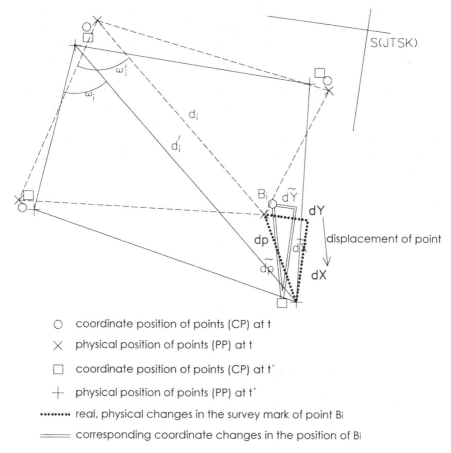

○   coordinate position of points (CP) at t

×   physical position of points (PP) at t

□   coordinate position of points (CP) at t′

+   physical position of points (PP) at t′

•••••••   real, physical changes in the survey mark of point B_i

═════   corresponding coordinate changes in the position of B_i

Fig. 3.4 Geometric relationship between the physical and coordinate positions of points and measured variables at epochs t and $t′$

which components (and their "generators") give rise to non-stochastic values and conversely, which components do not induce incompatible conditions.

Consider the scheme (Fig. 3.4) of surveyed and adjusted network structure of 4 points with physical position (PP) and coordinate position (CP) of its points.

Assume that the network was surveyed with a high quality and determined coordinates of points in the coordinate system used $S(XY) \therefore S(JTSK)$ correspond to the relevant PP within the limits of stochastic surveying tolerances (the effect of random errors in measurements).

Assume also that PP of 3 points did not change for the epoch $t′$ (eventually only insignificantly within the accuracy of their determination), except for one point B_i, for which there is shown that the apparent displacement of its physical mark (physical displacement of point) occurred at the time $t′ - t$. Let the points were determined in the datum $D(S(XY))$ at the epoch t, with measured variables L_i and with estimates of coordinates C_i and determined in a different datum $D'(S(XY)')$ for the epoch t' with measured variables L' and estimates of coordinates C'.

Assuming differences of measured elements relating to the changed point (e.g., due to movement of the physical mark of the point B_i):

$$dL = L' - L = [\ldots d_i' \ldots \omega_i' \ldots]^T - [\ldots d_i \ldots \omega_i \ldots]^T = [\ldots dd_i \ldots d\omega_i \ldots], \quad (3.10)$$

their causal analysis and changes in the configuration of point structure from the epoch t to t' imply that they particularly include:

- the influence of inaccuracy of measurement of elements of L at the epoch t and t',
- the influence of (unknown) displacement (shift) of the point's B_i physical mark, and
- the influence of various surveying technologies at t and t'.

Subsequently, using variables L, L', appropriate analytical methods, and their interaction, it is also possible to identify in some situations:

- points that did not change physically over time $t' - t$ (dL are small stochastic values at such points, generated only by inaccuracies of L and L' determination);
- points that are physically changed over time $t' - t$ (dL are significant values, generated not only by the inaccuracy of L, L' determination but also by an unknown size of the physical change of point).

When investigating the compatibility of geodetic control on the basis of the measured values L, L', values of $dd_i, \ldots d\omega_i, \ldots$ are not affected by coordinate determinations and consequently by corresponding datums, but only by measurement inaccuracies of geometric connecting elements between points.

In the examination of planimetric compatibility of points, if one will work with their unknown (determined) coordinates at epochs t and t', thus with coordinate indicators:

$$dC_i = C_i' - C_i = [XY]_i' - [XY]_i = [dX\,dY]_i, \quad (3.11)$$

analysis of their determination and results of changes incurred in a geodetic control indicates (Fig. 3.4) that dC are formed especially by superposition of effects:

- the influence of erroneous determination of points at the epoch t,
- the influence of erroneous determination of points at the epoch t',
- the influence of different datums at t and t',
- the influence of the unknown size of physical shift of (mark) point, and
- the influence of various surveying technologies at t and t'.

Therefore, in that case, when differences dC of coordinates C from epochs t and t' are used as indicators, they can be used directly for localization of:

- points that did not change over time $t' - t$ ($d\mathbf{C}$ are small values generated only by inaccuracies of measurement and determination of coordinates $\mathbf{C}, \mathbf{C}'$);
- points that are physically changed over time $t' - t$ ($d\mathbf{C}$ will represent non-stochastic values, generated not only by inaccuracies of measurement and determination of $\mathbf{C}, \mathbf{C}'$ but also by the unknown size of the physical change of point and other effects at t and t').

Various mathematical (geometric) relations between indicators $d\mathbf{L}$ and $d\mathbf{C}$, which may be exact or only approximate, can be used to their mutual conversion.

Values of indicators $d\mathbf{L}$ and $d\mathbf{C}$ examined from measurements and calculations represent only estimates of real and their unknown values $d\tilde{\mathbf{L}}, d\tilde{\mathbf{C}}$, and regarding the estimation theory, the larger and more accurate files will be used for the creation of estimates, the better the estimates will be.

3.4.4   Characteristics of Indicators

Both groups of indicators can be used to verify the compatibility of geodetic control, however, with different applicability regarding their capability, while $d\mathbf{L}$ indicators have fewer opportunities for usability.

In the past, although especially $d\mathbf{L}$ indicators were used of the both types of indicators, currently $d\mathbf{C}$ indicators or a combination of $d\mathbf{L}$ and $d\mathbf{C}$ is preferred so that disadvantages of $d\mathbf{L}$ are eliminated and conversely, that their complementary properties were applied in their use.

$d\mathbf{L}$ indicators (the change of geometric variables between points) have the advantage that they are independent (invariant) on the used coordinate system S (XY) and thus on the system datum $D(S(XY))$. It results from the properties of variables of $\mathbf{L}$, which are invariant elements to the system $S(XY)$, and therefore, $d\mathbf{L}$ derived therefrom have the same characteristics and usability, i.e., also $d\mathbf{L}$ indicators are independent of the coordinate system used.

The use of $d\mathbf{L}$ in analyses of compatibility of geodetic controls also has disadvantages. By using $d\mathbf{L}$ (without considering the relevant $d\mathbf{C}$ indicators), it is not possible to reliably or clearly determine oriented physical changes of points and their corresponding coordinate changes in space or coordinate system $S(XY)$, which are crucial characteristic signs of good or poor compatibility of the examined point and geodetic control. Besides, determination of the condition of points themselves is always the final objective of compatibility investigation.

For example, in the verification of compatibility between 3 points of geodetic control at the epoch t' with respect to the epoch t, Fig. 3.5 shows that physical survey marks of points j and k shifted to positions j', k' with significant values of dC_i, dC_k (Fig. 3.5a). In the verification of mutual geometric relations of points i, j, and k by measuring values of $\mathbf{L}$ only (some of distances and angles, e.g., $d_{ij}, d_{ik}, \omega_{jik}$), their changes may not indicate or also determine new changed physical positions of points, as shown in the scheme concerned. Also, the following

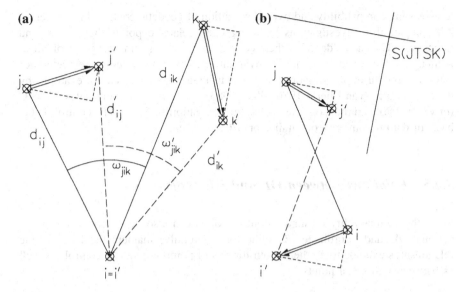

Fig. 3.5 The possible case of point incompatibility. **a** changes in physical and coordinate positions of points $X - X' = dX$, $Y - Y' = dY$ and **b** changes of certain $\mathbf{L}$ $\left(d_{ij} - d'_{ij} \nabla 0, \ \omega_{jik} - \omega'_{jik} \nabla 0 \right)$ not caused by them

relationships can be valid for elements $\mathbf{L}$ and $\mathbf{L'}$ after the change of physical points to positions j', k':

$$d_{ij} \cong d'_{ij}, \omega_{jik} \cong \omega'_{jik}, d_{ik} \neq d'_{ik}, \qquad (3.12)$$

therefore:

$$dd_{ij} \cong 0, d\omega_{jik} \cong 0, dd_{ik} \neq 0, \qquad (3.13)$$

while it is not possible to clearly assume the change of physical positions of survey marks of points j and k based on their values.

Similarly for the change of physical position of 2 points i and j (Fig. 3.5b), the distance d_{ij} valid at the epoch t can be also determined at the epoch t' by the size:

$$d'_{ij} \cong d_{ij}, dd_{ij} \cong 0, \qquad (3.14)$$

despite potential significant sizes of positional changes $d\mathbf{C}$ of end points i and j of this distance. Therefore, when using $d\mathbf{L}$ indicators, it is necessary to have available additional information about the geometrical situation in the geodetic control not only on the basis of $d\mathbf{L}$ but also on the basis of $d\mathbf{C}$.

Indicators $d\mathbf{C}$ are richer and more explicit in terms of their information content of the behavior and incorrect survey of points, and they are therefore currently used to assess the compatibility and overall condition of geodetic controls in preference. All compatibility investigations by using these relate to points themselves, and indicators $d\mathbf{C}$ exactly describe their conditions from a geometric point of view. Although values of $d\mathbf{C}$ may not be invariant to coordinate systems, their reference frames, and their implementations, this disadvantage can be solved, for example, by the S-transform (van Mierlo 1980; Illner 1983, and others) of coordinates $\mathbf{C}' \Rightarrow \mathbf{C}$ (or vice versa), and thus, one can assess the compatibility of geodetic control on the basis of datum homogeneous indicators $d\mathbf{C}$.

3.4.5   Relations Between DL and DC Indicators

Since the compatibility of survey control points can also be described by observational $d\mathbf{L}$ and coordinate $d\mathbf{C}$ indicators, naturally, mathematically definable relationships exist between them, both for sets of points—geodetic controls as well as between individual points.

3.4.5.1   Relations Between DL and DC in a Geodetic Control

Consider a geodetic control created at the epoch t, where values of geometric elements of $\mathbf{L}$ between points were obtained by measurements, and coordinates $\mathbf{C}$ of points in the system $S(XY)$ were obtained by the adjustment of the network structure of geodetic control.

Let this geodetic control be also surveyed at the epoch $t' > t$, where values of $\mathbf{L}'$ and $\mathbf{C}'$ were determined from measurements and their adjustment. The compatibility of points of the relevant geodetic control at the epoch t', if the appropriate data on the geodetic control from t and t' are available, is the main subject of interest.

Relations between $\mathbf{L}$ and $\mathbf{C}$ at the epoch t as well as between $\mathbf{L}'$ and $\mathbf{C}'$ at the epoch t' are expressed by well-known equations of the LSM adjustment of both surveys (e.g., by using the Gauss-Markov regular model—Koch 1988 and others):

$$\begin{aligned}
\mathbf{C} &= \mathbf{C}^{\mathrm{o}} + (\mathbf{A}^{\mathrm{T}}\mathbf{Q}_{\mathbf{L}}^{-1}\mathbf{A})^{-1}\mathbf{A}^{\mathrm{T}}\mathbf{Q}_{\mathbf{L}}^{-1}(\mathbf{L} - \mathbf{L}^{\mathrm{o}}), \\
\mathbf{C}' &= \mathbf{C}^{\mathrm{o}} + (\mathbf{A}'^{\mathrm{T}}\mathbf{Q}_{\mathbf{L}'}^{-1}\mathbf{A}')^{-1}\mathbf{A}'^{\mathrm{T}}\mathbf{Q}_{\mathbf{L}'}^{-1}(\mathbf{L}' - \mathbf{L}^{\mathrm{o}}),
\end{aligned} \tag{3.15}$$

where

$\mathbf{C}^{\mathrm{o}}$	approximate values of coordinates at both epochs,
$\mathbf{A}, \mathbf{A}'$	matrices of the configuration of the network structure of geodetic control at both epochs,
$\mathbf{Q}_{\mathbf{L}}, \mathbf{Q}_{\mathbf{L}'}$	matrices of cofactors of observed variables at both epochs, and
$\mathbf{L}^{\mathrm{o}} = f(\mathbf{C}^{\mathrm{o}}, \ldots)$	approximate values of observed variables.

In relations (3.15), the following relations can be labeled:

$$\mathbf{U} = \left(\mathbf{A}^T\mathbf{Q}_L^{-1}\mathbf{A}\right)^{-1}\mathbf{A}^T\mathbf{Q}_L^{-1},$$
$$\mathbf{U}' = \left(\mathbf{A}^T\mathbf{Q}_L^{-1}\mathbf{A}'\right)^{-1}\mathbf{A}'^T\mathbf{Q}_L^{-1}, \tag{3.16}$$

and subsequently, as defined for indicators $d\mathbf{C}$ (3.8):

$$d\mathbf{C} = \mathbf{C}' - \mathbf{C} = \mathbf{U}'(\mathbf{L}' - \mathbf{L}^\circ) - \mathbf{U}(\mathbf{L} - \mathbf{L}^\circ) \tag{3.17}$$

As follows from the Eq. (3.17), the linear expression $d\mathbf{C}$ at $d\mathbf{L}$, respectively, vice versa, is not exactly possible, and relations between $d\mathbf{C}$ and $d\mathbf{L}$ in geodetic control can be created only approximately based on different simplifications. However, in the global assessment of geodetic control, even such simplified approaches can provide applicable information about defects of incompatibility of points of geodetic control and their configuration, but they are not suitable for exact conversions between $d\mathbf{C}$ and $d\mathbf{L}$.

Of the possible approximate relationships between $d\mathbf{C}$ and $d\mathbf{L}$ in geodetic control, for example, the relation created on the following simplifications can be used for the overall average assessment of the compatibility of geodetic control:

– The same (at least approximately) structure (configuration) of the network at epochs t and t', formed by identical points and measured variables, was used in geodetic control, i.e., $\mathbf{A}' \cong \mathbf{A}$.
– Measurements were realized at epochs t and t' on the same level of accuracy, at least approximately, thus $\mathbf{Q}_{L'} \cong \mathbf{Q}_L$.

Subsequently, for (3.16) applies:

$$\mathbf{U} \cong \mathbf{U}' \tag{3.18}$$

And from Eq. (3.17) follows:

$$d\mathbf{C} \cong \mathbf{U}(\mathbf{L}' - \mathbf{L}) = \mathbf{U} \cdot d\mathbf{L} \tag{3.19}$$

Expression (3.19) indicates that the $d\mathbf{C}$ indicator is formed (approximately) by the projection of $d\mathbf{L}$, while the $\mathbf{U}$ projector expresses characteristics of both conditions of geodetic control at epochs t and t' as about equally to characterize.

A numerical identity applies to quadratic forms of coordinate and observational indicators $d\mathbf{C}$ and $d\mathbf{L}$, i.e., between:

$$\mathbf{R}_{dC} = d\mathbf{C}^T\mathbf{Q}_{dC}^{-1}d\mathbf{C},$$
$$\mathbf{R}_{dL} = d\mathbf{L}^T\mathbf{Q}_{dL}^{-1}d\mathbf{L}, \tag{3.20}$$

as pointed out in Sect. 6.3.

3.4.5.2   Relations Between DL and DC at Points

Indicators $d\mathbf{L}$ and $d\mathbf{C}$ between individual points of 2D geodetic control can be considered as differentials in the corresponding differential relations between points, resulting from the known definition equations between coordinates of points $\mathbf{C}$ and their geometric connecting elements $\mathbf{L}$.

For example, known relations apply between 3 points i, j, and k of geodetic control (Fig. 3.6) with coordinates $C_i = [X\,Y]_i$ (similar for C_j, C_k) and corresponding geometric connecting elements; thus, also the following relationship applies for the distance between points i and j:

$$d_{ij} = \sqrt{\left(X_j - X_i\right)^2 + \left(Y_j - Y_i\right)^2},\qquad(3.21)$$

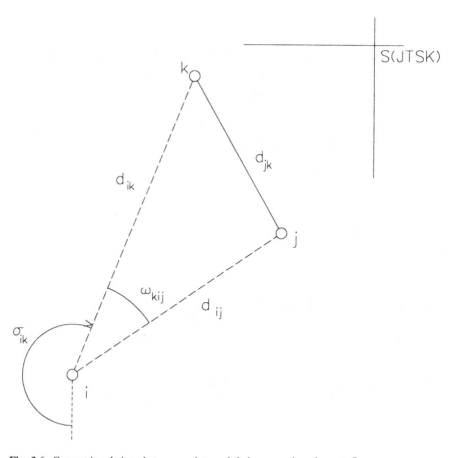

Fig. 3.6 Geometric relations between points and their connecting elements $\mathbf{L}$

and similarly for distances d_{jk}, d_{ki}; for included horizontal angles between points, for example, at the point i:

$$\omega_{kij} = \operatorname{arctg} \frac{Y_j - Y_i}{X_j - X_i} - \operatorname{arctg} \frac{Y_k - Y_i}{X_k - X_i}, \tag{3.22}$$

and similarly for other angles; to express coordinates as a function of known orientation (bearing σ_{ik}) and measured variables $d_{ij}, \omega_{kij}, \ldots,$, for example, for the point j, considering the points i and k:

$$C_j = \begin{bmatrix} X_j \\ Y_j \end{bmatrix} = \begin{bmatrix} X_i + d_{ij} \cos(\sigma_{ik} + \omega_{kij} \pm 200^g) \\ Y_i + d_{ij} \sin(\sigma_{ik} + \omega_{kij} \pm 200^g) \end{bmatrix}, \tag{3.23}$$

and similarly for the point j also considering other points.

Then, differential relations [total differentials of formulas (3.21)–(3.23)] apply for small changes of distances dd, angles $d\omega$ as well as coordinates of points $d\mathbf{C} = [dX \ dY]$:

$$dd_{ij} = \frac{\partial d_{ij}}{\partial X_i} dX_i + \frac{\partial d_{ij}}{\partial Y_i} dY_i + \frac{\partial d_{ij}}{\partial X_j} dX_j + \frac{\partial d_{ij}}{\partial Y_j} dY_j, \tag{3.24}$$

$$d\omega_{kij} = \frac{\partial \omega_{kij}}{\partial X_i} dX_i + \frac{\partial \omega_{kij}}{\partial Y_i} dY_i + \frac{\partial \omega_{kij}}{\partial X_j} dX_j + \frac{\partial \omega_{kij}}{\partial Y_j} dY_j + \frac{\partial \omega_{kij}}{\partial X_k} dX_k + \frac{\partial \omega_{kij}}{\partial Y_k} dY_k, \tag{3.25}$$

$$dX_i = \frac{\partial X_i}{\partial d_{ij}} dd_{ij} + \frac{\partial X_i}{\partial \omega_{kij}} d\omega_{kij},$$

$$dY_i = \frac{\partial Y_i}{\partial d_{ij}} dd_{ij} + \frac{\partial Y_i}{\partial \omega_{kij}} d\omega_{kij}, \tag{3.26}$$

that are specified by determination and substitution of the relevant partial derivatives. Thus, for example, for (3.24), after formulating these derivations:

$$\begin{aligned}
\frac{\partial d_{ij}}{\partial X_i} &= -\frac{\Delta X_{ij}}{d_{ij}}, \\
\frac{\partial d_{ij}}{\partial X_j} &= -\frac{\Delta Y_{ij}}{d_{ij}}, \\
\frac{\partial d_{ij}}{\partial Y_i} &= \frac{\Delta X_{ij}}{d_{ij}}, \\
\frac{\partial d_{ij}}{\partial Y_j} &= \frac{\Delta Y_{ij}}{d_{ij}},
\end{aligned} \tag{3.27}$$

for the differential dd_{ij}:

$$dd_{ij} = -\frac{\Delta X_{ij}}{d_{ij}} dX_i - \frac{\Delta Y_{ij}}{d_{ij}} dY_i + \frac{\Delta X_{ij}}{d_{ij}} dX_j + \frac{\Delta Y_{ij}}{d_{ij}} dY_j, \qquad (3.28)$$

or if derivations are expressed using the bearing of corresponding distance:

$$dd_{ij} = -dX_i \cos \sigma_{ij} - dY_i \sin \sigma_{ij} + dX_j \cos \sigma_{ij} + dY_j \sin \sigma_{ij}. \qquad (3.29)$$

Similar relations can also be specified for other differentials (3.25) and (3.26).

The above relations can be used for various partial analyzes of the mutual influence of $d\mathbf{L}$ and $d\mathbf{C}$ (mostly by individual points), in which, however, the knowledge about changes of certain variables or coordinates in terms of the effect of the whole geodetic control cannot be obtained. For example, dC_i according to (3.26) should be determined on the basis of measured variables (d, ω) and their determined changes $(dd, d\omega)$ not only with respect to $dd_{ij}, d\omega_{kij}$ but also considering other distances $d_{ik}, \ldots$ directed from other points of geodetic control to the point i. It is evident that these values will be different for each partial determination of dC_i and definitive changes of dC_i cannot be clearly determined from them by any geometric solution. Only stochastic significances of dC_i can be assessed by statistical methods.

Therefore, vectors $d\mathbf{C}$ of indicators $d\mathbf{L}$ and $d\mathbf{C}$ will be preferred as dominant indicators for monitored objectives in further analyses of compatibility, since these data are unique in each geometric situation and are directly related to the point (coordinates), whose characteristic, compatibility, is examined.

3.5   Pre-information to Verify Planimetric Compatibility

Data on the survey and adjustment of the network structure of geodetic control in both epochs t and t' are necessary to verify the compatibility of points of geodetic control at epoch t'. Therefore, to examine the compatibility of points, one need not only the current data—information about their condition from the present epoch t' —but also about their condition in the previous epoch t, in which the geodetic control was surveyed and adjusted. It is required by the very principle of verifying the compatibility of 2D points (Chap. 6), which is based on the comparison of two states of the same geodetic control at different epochs t and t' and in the determination of changes (in connecting elements, coordinates of points) at the epoch t' against the state at the epoch t. Consequently, one need to know mainly the following information—data on the examined geodetic control and its network structure, both from the epoch t and the epoch t':

$\mathbf{C} = [X\,Y], \mathbf{C}' = [X'\,Y']$ coordinates of points of geodetic control at t and t',
$\Sigma_\mathbf{C}, \Sigma_{\mathbf{C}'}$ covariance matrices of coordinates $\mathbf{C}, \mathbf{C}'$,

$s_0^2, s_0'^2$	a posteriori unit variances from the adjustment of the network structures at epochs t and t',
$\mathbf{L}, \mathbf{L}'$	vectors of measured variables at t and t',
$\mathbf{Q_L}, \mathbf{Q_{L'}}$	cofactor matrices of variables $\mathbf{L}, \mathbf{L}'$,
$\mathbf{v}, \mathbf{v}'$	vectors of corrections (of measured variables) from both adjustments, data on the extent of measurements, redundancy, etc.,

and also other, more detailed information in specific cases of verification of compatibility of geodetic controls.

On the basis of disposition of the necessary information for the realization of compatibility survey (according to their availability), two different information situations (models) are formed for the compatibility verification:

(a) All of the above, necessary data from the old epoch t, i.e., the complete documentation of survey and adjustment of the examined geodetic control, are available;

as well as the entire documentation of the present survey and processing of the relevant geodetic control of the current epoch t' is also available.

(b) Only the coordinates of points $\mathbf{C}$ of the relevant geodetic control and no more data are available from the old epoch t; all necessary data from the measurements and adjustment are available from the current epoch t'.

The pre-information situation (b) apparently relates to an earlier, previously established networks on the national territories, in various regions, etc., of which, however, the documentation and relevant data were not retained to the extent necessary (previously, many of the necessary data have not been even determined in the establishment of networks), and only coordinates of points $\mathbf{C}$ are available, without the possibility of an objective assessment of their accuracy, reliability, and other characteristics.

It is obvious that it will be necessary to use various strategies and procedures in the examination of compatibility of points for these different situations. They will be discussed in Chaps. 6 and 7.

Chapter 4
The Compatibility of 1D (Height) Points

In the issue of height compatibility of different types of vertical controls (levelling, trigonometric, GPS, etc.), only levelling point fields, respectively, networks, that are so far dominant not only for the vertical expression of points relative to the geoid, or quasigeoid, but also in terms of the technology and accuracy of height determination, will be considered. Therefore, reflections on the compatibility of height points, its determination, and analysis will refer to leveling points (networks) and scalar parameters (height data) attributed to them. Formally, the procedures are also applicable for vertical networks surveyed by other technologies, for example trigonometrically.

4.1 Characteristics of Height Points in Terms of Compatibility

In terms of construction and monumentation on the surface, each height point HL is composed of a point bearer and height survey mark itself (Fig. 4.1), to which a hypothetical horizontal tangent plane forming the point of tangency T (for a leveling rod) applies, realizing the given height point. Also, the tangent level surface of the field of gravity with the height data h should pass through the point T, i.e., both surfaces should be stochastically identical at the point T. Then, the height situation at the point T is eligible for the use of point also from a theoretical perspective and in this case, the point can be declared as a compatible height point.

Otherwise, if a non-random, vertical difference that is significant from a statistical point of view exists between the tangent plane and the level surface, the height point represents an incompatible height point that is not suitable for geodetic activities.

© The Author(s) 2016
G. Weiss et al., *Survey Control Points*,
SpringerBriefs in Geography, DOI 10.1007/978-3-319-28457-6_4

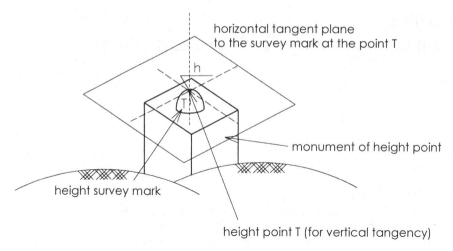

horizontal tangent plane
to the survey mark at the point T

h

T

monument of height point

height survey mark

height point T (for vertical tangency)

Fig. 4.1 The height point HP and its two components (survey mark T, height data h)

4.2 The Emergence of Incompatibility and Its Effect on Height Determination

The incompatibility of height point HL, similar to planimetric points (Sect. 3.2), may occur in relation to a certain period $t' - t$ for the following reasons:

- a height survey mark with a bearer may shift spatially at the time $t' - t$ due to the action of various forces on the mass around the point, while the vertical component of shift is caused by height change of survey mark (of the point T) with a tangent plane,
- height data h of the point were incorrectly determined (due to a hidden, undiscovered error in measurements or calculations, or due to the use of non-identical vertical datums) during its establishment (at the epoch t),

It is evident, even without detailed analyzes, that in the case of height determination of new points (or heights control of given points) grouped into a network structure, in which height points—connecting (datum) HL and also an incompatible point among them will exist, this incompatible point will cause significantly incorrect data for determined or controlled heights at some points.

4.3 Indicators of Height Incompatibility and Their Characteristics

Even in the present case, the compatibility of height points HL is verified on the basis of:

– their known heights h from the epoch t (from the period of the initial establishment and survey of points),
– determined heights h' from the current epoch t'.

Comparison of both data h and h' at each point P will provide a height difference —height discrepancy:

$$V_h \equiv dh = h' - h, \tag{4.1}$$

that is used as an indicator of height compatibility of the corresponding point H. In terms of general principles of statistical assessment of the significance of random variables (Hald 1972; Anděl 1972; Radouch 1983 and others), if V_h is insignificant, the stochastic value from a numerical perspective and the difference of heights h, h' will be considered as a random value, indicating that the relevant point can be considered as compatible. Conversely, if V_h will be of significantly large, unacceptable value for a given point, it will be a non-stochastic, statistically significant size of the change of point height and this point will be considered as the point incompatible in height. The subjectivity of the assessment of the V_h size, also in the present case, is eliminated using the appropriate statistical verification, according to which the tested point with its discrepancy is included among compatible or incompatible height points of the examined geodetic control.

In common practice, only the V_h indicator (analogy of V_X, V_Y for planimetric points) is used to assess the height compatibility of points.

4.4 Pre-information to Verify the Height Compatibility

Analogously to planimetric geodetic controls, the examination of compatibility of vertical controls requires various data and information relating to two different time measurements and adjustments of vertical control.

From the epoch t (the time of establishment of geodetic height points, their measurement and adjustment), all data and technical pre-information relating in particular to the results of determination of heights and their characteristics are required. These are mainly:

– the adjusted height of points,
– their cofactor matrix,
– a posteriori unit variance,
– configuration characteristics of the network and its survey,
– and possibly additional information.

The same information as for the epoch t is required at the epoch t', in which the current survey of the relevant height points and also the adjustment of their network structure are realized.

However, the indicated data for the examination of compatibility of height points from both epochs may not always be available to the extent necessary. According to the content of pre-information that we actually have for the height verification of points, two characteristic different situations (models) also originate in the leveling geodetic controls:

(a) all the necessary data from the survey and adjustment of height points from the epoch t as well as from the current epoch t' are available,

(b) from the epoch t, only the heights h of leveling points HL in the corresponding space are available for the verification procedures, while from the current epoch t', all the necessary data and results of measurement and adjustment of height points are usable.

In each of those two situations, it is therefore necessary, as in 2D geodetic controls, to use various methods of compatibility verification, especially in the testing phase, of which the following will be specified in the next sections: the method of height congruency (Niemeier 1979, 1980) for cases with full pre-information and the Lenzmann–Heck method (Lenzmann 1984; Heck 1985) for height structures with incomplete pre-information.

Chapter 5
The Compatibility of 3D Spatial Points

By using the GNSS technology (GPS, GLONASS, …), spatial positions of points in 3D coordinate systems (WGS-84, ETRS-89, …) are determined. As required by their geodetic use, they may or may not be transformed into national planimetric or/and vertical coordinate systems. Afterward, also the quality (or stability) of determined points is assessed separately in them.

At present, when European countries have established their spatial networks in the ETRS-89 system, which may also be used for local geodetic activities (for example, repeated measurements and 3D determinations of points of geodetic networks), also the compatibility issues (or changes between epochs) and well-researched decisions on 3D stability—compatibility (or a spatial change of examined points), are up-to-date.

5.1 Characteristics of Spatial Points in Terms of Compatibility

Every spatial point, in terms of its construction and monumentation on the earth surface, consists of a point bearer and the spatial survey mark itself. Consider defined spatial positions of points $C_i = [X_i, Y_i, Z_i]$, where $i = 1, 2, … p$, in a specific spatial system $S(XYZ)$ (Fig. 5.1).

The functionality of spatial point is conditioned by three essential components:

- physical position is bound to the earth surface, or objects attached to the surface and is variable in relation to close or distant surroundings as a result of various forces acting on the point bearer with the spatial survey mark,
- spatial coordinates $C_i = [X_i, Y_i, Z_i]$ of physical mark of point in a certain spatial system $S(XYZ)$ that allow various computations, geodetic activities, and field survey by using a survey control point. Under the influence of additional corrections of global or local spatial systems of the given country, also the coordinate position of a point can change, which will result in the coordinate instability of the point in time.

© The Author(s) 2016
G. Weiss et al., *Survey Control Points*,
SpringerBriefs in Geography, DOI 10.1007/978-3-319-28457-6_5

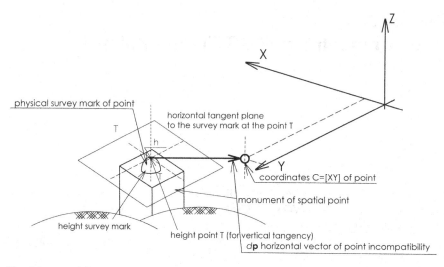

Fig. 5.1 A spatial survey point and its components (survey mark, spatial coordinates)

In order that the spatial point C_i is compatible (in terms of quality characteristics), its two components must permanently correspond to each other, i.e., its coordinate and physical survey mark have a stochastic spatial identity in the used spatial system. Alternatively, the physical survey mark of a spatial point corresponds to the relevant coordinates in the 3D system $S(XYZ)$.

Compatible spatial points in particular space may form a compatible geodetic control. If a geodetic control contains a certain number of incompatible spatial points, thus it may be a moderate, strong, etc. incompatible geodetic control.

For spatial coordinate systems, it is appropriate to consider what type of coordinate system will be selected. As already stated in the previous chapters on changes in position or height in national coordinate systems, a change in one coordinate of position or a change in height will be reflected as the change of all three coordinates (Fig. 5.2) in spatial coordinate systems. In national coordinate systems, one is usually accustomed that spatial placement of a point is divided into planimetric and vertical placement, in order to make clear that a spatial point is incompatible due to the planimetric or vertical displacement of the point.

Where

$$\Delta h = \sqrt{\Delta X^2 + \Delta Y^2 + \Delta Z^2}.$$

In such a case, a network of compatible spatial points can be divided into the network of points compatible in position and the network of points compatible in height in national coordinate systems of the countries concerned (Sects. 3.1 and 3.4).

Fig. 5.2 The altimetric change in 1D will be reflected by the changes in 3D in all coordinates ΔX_i, ΔY_i, ΔZ_i

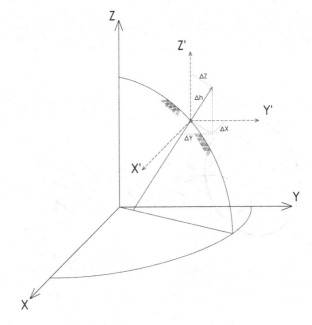

5.2 The Emergence of Incompatibility of Points and Its Effect on the Establishment of Spatial Networks

If incompatible points are located in certain spatial geodetic control, it is necessary to identify them in order to increase the quality of the geodetic control as a whole. Moreover, in the case of using the geodetic control for geodetic activities and calculations, exclude incompatible spatial points from it and do not use it for the above purpose.

Incompatible geodetic controls usually originate due to the stage completion or expansion of geodetic control in areas where the insufficient density of points, from which it is necessary to realize geodetic activities or calculations, is present. The verification of compatibility of a spatial geodetic control represents the verification of determination of 3D coordinates of points and their time stability, or instability since the compatibility is still related to a certain time period.

The emergence of incompatible points in the establishment of spatial geodetic controls can be visualized according to Fig. 5.3, where also the Z-coordinate in the spatial coordinate system will be considered in the stage expansion of a spatial geodetic control.

However, spatial displacements will represent not only positional changes in the XY plane of given spatial coordinate system but also height changes in Z-coordinate. The principle of stage establishment of geodetic controls and the emergence of incompatible points is explained in Sect. 3.2.

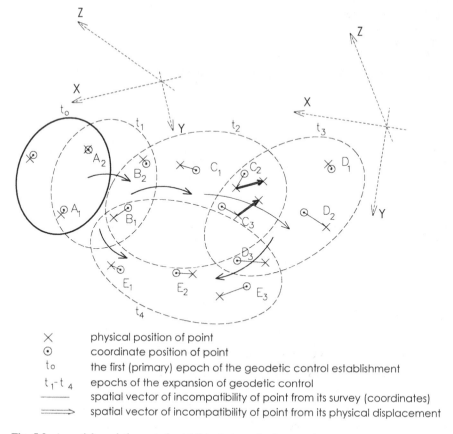

$\times$	physical position of point
$\odot$	coordinate position of point
t_0	the first (primary) epoch of the geodetic control establishment
t_1-t_4	epochs of the expansion of geodetic control
———	spatial vector of incompatibility of point from its survey (coordinates)
===>	spatial vector of incompatibility of point from its physical displacement

Fig. 5.3 A spatial geodetic control established at epochs (t_0, ..., t_4)

A geodetic network is a structural formation on and below the surface, consisting of a set of physical points, to which certain parameters are assigned. A geodetic network consists of three basic components: the geodetic control of survey control points appropriately distributed and physically monumented, and eventually permanently targeted in terrain; geometric connecting elements, i.e., sets of measurable geometric and physical parameters between points, for example spatial distances, observed directions, zenithal distances, height differences; the datum of geodetic control defining the system, in which network points have assigned coordinates and other parameters. Equivalent mathematical relationships apply between compatible and incompatible points and corresponding geometric connecting elements. Among incompatible points of a spatial network, the relevant geometric elements will have significantly changed values.

The incompatibility of spatial points is caused, as it has already been characterized in Sects. 3.2 and 4.2, by the following:

- the above limit inaccuracy of coordinate determination of a physical survey mark (as a result of measurement and processing itself, nonidentical datum of points, but also as a result of used surveying instruments),
- spatial change of a physical survey mark of point (displacement, subsidence) caused by different deformation forces acting on the physical bearer of survey control point.

5.3   Indicators of Spatial Compatibility of Points and Their Characteristics

5.3.1   In General

As already mentioned in the previous chapters for the verification of compatibility in a planimetric or vertical geodetic control, it is also necessary to have the following information and data for the verification of compatibility of a spatial geodetic control:

- from spatial measurements and processing at two different epochs t and t',
- or from the documentation of the previous epoch t and spatial measurements and processing at the current epoch t'.

The compatibility of points is always determined for the time t' considering their condition at the time t based on the comparison of:

- values of spatial geometric connecting elements of a geodetic network's control (spatial distances d, observed directions (angles) ω, zenithal distance z, height differences Δh, coordinate differences of vectors ΔX, ΔY, ΔZ from observations by GNSS technology (Labant et al. 2011), etc.) between epochs t and t' in order to determine differences of measured values—observational indicators:

$$d\mathbf{L} = \mathbf{L} - \mathbf{L'},$$

where
$$\mathbf{L} = \{d, \omega, z, \Delta h, \Delta X, \Delta Y, \Delta Z, \ldots\},$$
$$\mathbf{L'} = \{d', \omega', z', \Delta h', \Delta X', \Delta Y', \Delta Z', \ldots\},$$

- values of determined coordinates of points at epochs t a t', in order to determine differences of coordinates—coordinate indicators:

$$d\mathbf{C} = \mathbf{C} - \mathbf{C'},$$

where
$$\mathbf{C} = f(\mathbf{L} = \{d, \omega, z, \Delta h, \Delta X, \Delta Y, \Delta Z, \ldots\}),$$
$$\mathbf{C'} = f(\mathbf{L'} = \{d', \omega', z', \Delta h', \Delta X', \Delta Y', \Delta Z', \ldots\}),$$

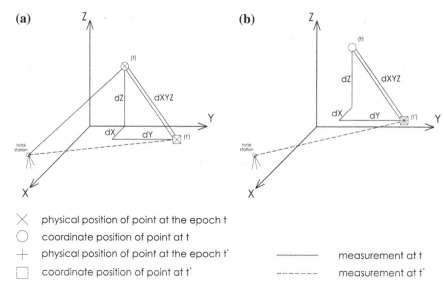

(a)

(b)

$\times$ physical position of point at the epoch t

$\bigcirc$ coordinate position of point at t

$+$ physical position of point at the epoch t'   ———————   measurement at t

$\square$ coordinate position of point at t'   - - - - - -   measurement at t'

Fig. 5.4 Graphical representation of *dXYZ* (*dX*, *dY*, *dZ*) in both types of incompatibility

By using mathematical processing and testing of statistical hypothesis, one can determine compatible points whose physical and coordinate positions in the space have not changed over time $t' - t$; or incompatible points whose physical survey mark of point has changed by the vector *dXYZ* (Fig. 5.4a) or they were determined by incorrect coordinates (the most commonly in the previous epoch) (Fig. 5.4b).

Based on the values of observational indicators *d***L** or coordinate indicators *d***C**, the compatibility, or incompatibility, of point (geodetic control) can be demonstrated. Small values of indicators indicate the compatibility of point (statistically *dXYZ* = 0) while high values of *d***L** and *d***C** indicate the incompatibility of point (statistically *dXYZ* ≠ 0). The dividing line between compatibility and incompatibility is determined by suitable mathematical methods of statistical hypothesis testing.

5.3.2   Types of Indicators

The creation of indicators such as *d***L** and *d***C** is based on the measurements of geometric connecting elements in a geodetic network at the time *t* and *t'*, while information about the compatibility of geodetic control will be at the time *t'*. The observational vector **L** of measured geometrical values with the covariance matrix of measured values will be created at the epoch *t* after the measurements:

$$\mathbf{L} = (d_1, d_2, \ldots \omega_1, \omega_2, \ldots z_1, z_2, \ldots \Delta h_1, \Delta h_2, \ldots \Delta X_1, \Delta X_2, \ldots \Delta Y_1, \Delta Y_2, \ldots \Delta Z_1, \Delta Z_2, \ldots)^{\mathrm{T}}, \mathbf{\Sigma_L}.$$

After the processing of observational vector obtained at p determined points, the estimates of coordinates $\hat{\mathbf{C}}$ with the covariance matrix of coordinate estimates will be obtained:

$$\hat{\mathbf{C}} = (X_1, Y_1, Z_1, \ldots X_p, Y_p, Z_p)^{\mathrm{T}}, \mathbf{\Sigma_C}.$$

At the epoch t', the observational vector $\mathbf{L'}$ with $\mathbf{\Sigma_{L'}}$ and estimates of adjusted coordinates with $\mathbf{\Sigma_{\hat{C}'}}$ will be defined after the survey of geodetic network:

$$\mathbf{L'} = (d_1', d_2', \ldots \omega_1', \omega_2', \ldots z_1', z_2', \ldots \Delta h_1', \Delta h_2', \ldots \Delta X_1', \Delta X_2', \ldots \Delta Y_1', \Delta Y_2', \ldots \Delta Z_1', \Delta Z_2', \ldots)^{\mathrm{T}}, \mathbf{\Sigma_{L'}}$$
$$\hat{\mathbf{C}}' = (X_1', Y_1', Z_1', \ldots X_p', Y_p', Z_p')^{\mathrm{T}}, \mathbf{\Sigma_{C'}}.$$

Based on differences of measured variables $d\mathbf{L} = \mathbf{L'} - \mathbf{L}$ and differences of estimates of coordinates $d\mathbf{C} = \mathbf{C'} - \mathbf{C}$, it is possible to assess the compatibility of points of the geodetic network.

Indicators $d\mathbf{L}$ and $d\mathbf{C}$ are small values if the stability of physical survey marks of points or determination of points was realized from measured variables containing no outlying (biased) values. Therefore, the geodetic control and all of its points can be considered as compatible and there were no significant changes in their coordinate and physical component.

Indicators such as $d\mathbf{L}$ and $d\mathbf{C}$ also contain high values (statistically significant values) if the instability of physical survey marks of points, or determination of points, was realized from measured variables containing outlying (biased) values. Therefore, the geodetic control and one/some of its point/s can be considered as incompatible, so significant changes in its/their coordinate and physical component have occurred at this(these) point(s).

Based on the size of indicators $d\mathbf{L}$ or $d\mathbf{C}$, it is possible to determine the spatial vector of incompatibility at incompatible points (or all points):

$$dXYZ = \left(\begin{array}{c} \vdots \\ dXYZ_i = \sqrt{dX_i^2 + dY_i^2 + dZ_i^2} \\ \vdots \end{array} \right),$$

which represents spatial relations between epochs t and t'.

Chapter 6
The Verification of Compatibility of Planimetric Points

6.1 In General

The verification of compatibility of points in planimetric geodetic controls that are part of national, regional network structures, or individual local, purpose-built networks represents a very important component of all processes of complementing and expansion of networks by new points that are necessary for different geodetic activities, requiring reliable geodetic bases, i.e., points. This reliability cannot be achieved only by a quality survey of new points, since even the quality of used old points as connecting (datum) points in terms of their compatibility, or compatibility of the whole relevant geodetic control, is no less important aspect of this task.

A variety of suitable methods have been developed for the verification of compatibility of planimetric points, despite the difficulties that result from the lack of pre-information (Sect. 4.4). Those methods, with an acceptable credibility, verify the quality of existing points and geodetic controls in terms of their compatibility and, therefore, allow us to use them for various challenging geodetic tasks (establishment of new points, setting-out tasks, deformation monitoring, etc.).

The following chapter discusses various methods on how to determine compatibility or incompatibility of geodetic controls and how to "rectify" an incompatible geodetic control in order to be usable, regarding both information models that are typical for current geodetic controls and also in terms of the use of both types of indicators $d\mathrm{C}$ and $d\mathrm{L}$.

6.2 The Compatibility Verification with $d\mathrm{C}$ Indicators

Within the compatibility verification of group of existing points from a certain superior network or various individual networks that should be used for actual geodetic activities and tasks, one can proceed either by using $d\mathrm{C}$ indicators or

© The Author(s) 2016
G. Weiss et al., *Survey Control Points*,
SpringerBriefs in Geography, DOI 10.1007/978-3-319-28457-6_6

sometimes by using $d\mathbf{L}$ indicators. Recently, coordinate indicators are preferred, especially for the explicitness of characterization and expression of the physical and coordinate change of points, their theory and applicability is more sophisticated than in the case of $d\mathbf{L}$ indicators and also of the reasons that $d\mathbf{C}$ indicators characterize the compatibility state of points of every geodetic control directly and with full information. $d\mathbf{L}$ indicators are mainly used together with $d\mathbf{C}$ for additional, eventually verification characterization of compatibility of geodetic controls.

6.2.1 Strategies of Compatibility Verification

6.2.1.1 Verification of Compatibility with Full Pre-information

The model situation for these cases of compatibility verification is shown in Fig. 6.1. Points P form points of the superior network (e.g., ŠPS in a given area, of which p points were used as connecting points labeled PL to determine points B of local network LN at the epoch t. These points B and PL form an individual local network LN in a certain area (e.g., in the construction of various technical and industrial objects). The necessary documentation and data (Sect. 3.5) are available for the network LN:

$$\mathbf{C}, \boldsymbol{\Sigma}_{\mathbf{C}}, s_0^2, \mathbf{L}, \mathbf{Q}_{\mathbf{L}}, \mathbf{V}, \ldots \qquad (6.1)$$

from survey and adjustment of the network at the epoch t.

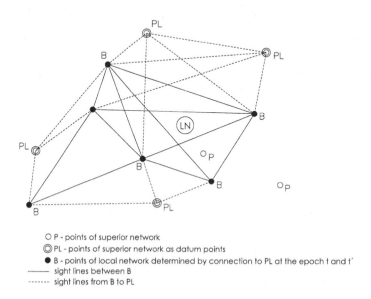

O P - points of superior network
◎ PL - points of superior network as datum points
● B - points of local network determined by connection to PL at the epoch t and t′
——— sight lines between B
------- sight lines from B to PL

Fig. 6.1 Local planimetric network LN of points B, PL, surveyed at epochs t and t'

The main objective is to examine the current compatibility of points B and PL in the area of LN, for the epoch t', since the LN in this area will be densified with the connection of new points to points B and PL, respectively, various geodetic activities will be realized using these points. The LN network has its parameters bound to the epoch t, at which it was established and to which all information and data are related (5.1).

To examine the compatibility of points of the network structure of LN, these points are also surveyed at the epoch t' (using practically the same configuration as at time t, at the same or even higher level of accuracy, using the same datum points, etc., if possible) and adjusted (the same method of adjustment, the same level of significance, etc.). Thus, the following data are obtained for points B and PL of the network LN from the epoch t':

$$\mathbf{C}', \Sigma_{\mathbf{C}'}, s_0'^2, \mathbf{L}', \mathbf{Q}_{\mathbf{L}}', \mathbf{V}', \ldots \tag{6.2}$$

as well as other parametric information. Thus, the pre-information situation (a) (Sect. 3.5) arises for the problem solution.

Based on the values, (6.1), (6.2), and other variables derived therefrom, mainly discrepancies dC and their accuracy, they are verified by appropriate testing procedure, whether they indicate compatibility or incompatibility of points from the LN for the epoch t'.

If the method of verification confirms that coordinates $\mathbf{C}, \mathbf{C}'$ of all points B and PL indicate compatibility of points, i.e., significant differences do not appear between $\mathbf{C}$ and $\mathbf{C}'$, all points in the LN can be used for the planned, current geodetic activities in a given area.

However, if the method of verification proves that the coordinate indicator dC has a significant value between coordinates $\mathbf{C}$ and $\mathbf{C}'$ of some point (of PL or B points), i.e., such a point is incompatible, that point will not be used for current activities. We assume that after the elimination of incompatible point and related geometric elements, the remaining geometric elements in the vector $\mathbf{L}$ will suit for the next new adjustment of LN in every respect (redundancy, configuration defect, etc.).

The new adjustment of the LN will be realized with the remaining $p - 1$ points PL and B using the corresponding geometric elements. We expect that after obtaining the new coordinates $\mathbf{C}'$ at this stage of the solution, their analysis together with $\mathbf{C}$ for the remaining PL and B points either confirms their compatibility or not. If the compatibility of geodetic control of the LN is not confirmed, i.e., if even one another point among points $p - 1$ is incompatible, the whole verification process is performed again, or it is repeated until the planimetric compatibility is proven for all points PL and B from the LN.

6.2.1.2   Verification of Compatibility with Incomplete Pre-information

Very commonly, only the coordinates $\mathbf{C} = [X\,Y]$ in a certain 2D coordinate system $S(XY)$, for example, S-JTSK, are available for the compatibility examination of the set of points from the epoch t. Thus, it is a pre-information situation (b) (Sect. 3.5), whose model structure can be illustrated by the situation in Fig. 6.2.

In a given area, points P are points of a superior network in the system S(JTSK), with known coordinates $\mathbf{C}$ from the epoch t. Points PL (with the number of points p) are also points of a superior network; they are used as connecting (datum) points to determine new points U (with the number of points u) at the epoch t'. Only the coordinates $\mathbf{C}$ are available for points P and PL; other necessary data on their survey and adjustment from the epoch t are unknown.

For points U from the epoch t', all necessary data from the survey of their network structure of LN are available, from its adjustment in a selected local system S(LOC) as well as from the transformation of these coordinates $\mathbf{C}^{L'}$ into the system S(JTSK) of superior network to coordinates $\mathbf{C}^t$ labeled as $\mathbf{C}^{Jt}$, while points PL represent identical points.

The quality and compatibility of selected points PL is an important requirement and condition of connecting points U to the superior network. Therefore, the task is to verify the compatibility of points PL, for which two groups of coordinates will result from the survey and adjustment of local network LN: given coordinates $\mathbf{C}$ from the epoch t and obtained coordinates $\mathbf{C}^t$ from the epoch t', those allow us to assess the condition of their identity—according to their statistical comparison.

Some of the known procedures (Sect. 6.2.2.1) are used to examine the compatibility of points PL.

Mostly, the algorithm that is based on the implementation of the following strategy is applied. Points PL, U are accurately surveyed at the epoch t', their

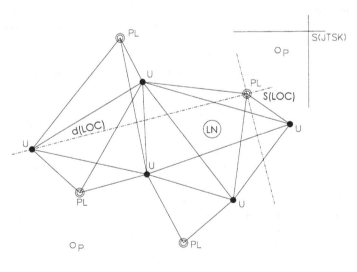

O P - points of superior network
◎ PL - poinst of superior network as datum points
● U - determined points in the LN

Fig. 6.2 Local planimetric network LN of points U, PL, determined at the epoch t in the S(LOC)

adjustment (most commonly with the model: Gauss–Markov singular—GMM with incomplete rank, Koch 1988) is realized in the coordinate system S(LOC) established by their contractor of work and in an appropriate computing plane (in our situation, most commonly in the plane of S-JTSK). From the obtained parameters:

$$\mathbf{C}_P^{\prime \mathbf{L}}, \mathbf{C}_U^{\prime \mathbf{L}}, \boldsymbol{\Sigma}_{\mathbf{C}'(P,U)}, s_0^{\prime 2}, \mathbf{L}', \mathbf{Q}_{\mathbf{L}'}, \mathbf{V}', \tag{6.3}$$

coordinates are transformed to $\mathbf{C}^t$ coordinates of points PL, U in the system S (JTSK) by the Helmert transformation (in the creation of transformation parameters of all points PL). Thus, we have two sets of coordinates for points PL in the S (JTSK): $\mathbf{C}$ from the epoch t and $\mathbf{C}^t$ from the current solution at the epoch t', which form coordinate indicators $d\mathbf{C} = \mathbf{C} - \mathbf{C}^t$ for points PL, on the basis of which the compatibility of every point PL is examined by appropriate methods of statistical testing for the epoch t'.

If the verification confirms the acceptable compatibility of points PL, an opinion can be adopted that the connection of points U to the superior network using connecting points PL is correct without significant disruptive effects of the surrounding geodetic control with points P and PL on determined coordinates of the group of new points U and those points PL, U in the area of the works form a compatible geodetic control, in which all geodetic activities can be realized.

If the verification of compatibility state of points PL proves the incompatibility for any of them, such a point is excluded from the set of points PL in this case. The process of transformation and interpretation is realized again with the remaining $p - 1$ points, and the compatibility of $p - 1$ points PL is verified. In the case of confirmation of their compatibility, results of the solution from this state are definitive. If an incompatible point occurs even among $p - 1$ points PL, the process is repeated until the state of compatibility of all remaining PL points.

Also another algorithm (combinative) can be used to verify the compatibility of points PL, without the transformation procedure, for situation models according to Fig. 6.2. Within this procedure, several combinations are created for datums from a sufficient number of points PL (3 or more points are used in each combination) and the LN network is adjusted with each combination. If all datum points are compatible, coordinates of points U (different from each other only at a stochastic level) are obtained from each combination. If any datum point PL is incompatible, it will be reflected in significantly different values of coordinates of some points U. Subsequently, it is possible to examine and identify those points PL that demonstrate the incompatibility and which caused incorrect coordinates of points U in the appropriate combinations using suitable logical mathematical procedures.

From both algorithms, the priority should be given to the transformation process, since the combinational procedure has apparent disadvantages:

- it is lengthier,
- incompatible points are determined by an indirect identification,
- by its principle, it is less sensitive to identify incompatible points as the transformation process.

6.2.2 Methods of Compatibility Verification

6.2.2.1 Overview of Methods

Progressively, with the growth in the importance of separation of "good" survey control points from inappropriate points in geodetic controls, several methods for the identification of "bad," incompatible points, i.e., for the verification of compatibility of network structure, were created. Various types of adjustment of measurements, transformation procedures, and various realizations of testing procedures are conveniently combined in them.

Essentially, each method can handle the required task of identification of incompatible points with the application of several common elements. These elements represent principles of task solution, and the following are the most important among them:

- coordinates $\mathbf{C}$ of verified points from their initial determination (the epoch t and earlier) are known,
- coordinates $\mathbf{C}'$ of verified points at present (the epoch t') are determined: thus, two measurements of different time and their results are necessary for the verification of points,
- the comparison of $\mathbf{C}$ and $\mathbf{C}'$ by the creation of their differences $\cdot\equiv$ coordinate indicators $\equiv$ coordinate discrepancies:

$$\mathbf{V_X} = d\mathbf{X} = X' - X, \quad \mathbf{V_Y} = d\mathbf{Y} = Y' - Y \qquad (6.4)$$

from which various discrepancy functions may be created, for example:

- $dp_i = \sqrt{V_{Xi}^2 + V_{Yi}^2}$, planimetric discrepancy—the vector of point incompatibility (Sect. 3.4.2),
- $dp_i^2 = V_{Xi}^2 + V_{Yi}^2$, quadratic planimetric discrepancy,
- $\mathbf{V}_i^T \mathbf{Q}_{Vi}^{-1} \mathbf{V}_i$, resp. $\mathbf{V}_i^T \mathbf{V}_i$, quadratic form of planimetric point discrepancy, where $\mathbf{V_i} = [dX_i \ dY_i]$,
- $\mathbf{V}^T \mathbf{Q}_V^{-1} \mathbf{V}$, resp. $\mathbf{V}^T \mathbf{V}$ quadratic form of discrepancies in the whole geodetic control, where $\mathbf{V} = [dX_1 \ dY_1 \ \dots \ dX_p \ dY_p]$ and $\mathbf{Q}_{V_i}, \mathbf{Q}_V$ are corresponding cofactor matrices of discrepancies—corrections (residuals).

On the basis of the general statistical concept of assessment of condition, character, and characteristics of phenomena, also discrepancies are random, necessary attributes of two measurements different over time. They can form small values of stochastic character (that are expected for the compatibility of points and their geodetic control), or even higher values, statistically significant, which indicate especially these sources of emergence of discrepancies:

- the physical position of the point has changed in the period $t' - t$,
- an unacceptable effect of different realizations of coordinate frame occurred as a result of the use of different datums,
- coordinates C at the epoch t were determined incorrectly (most commonly),
- coordinates C' at the epoch t' were determined incorrectly if statistical verifications of measured variables were not realized up to the mark,
- various surveying technologies without the elimination of their specific impacts were used.

Generally, the current measurements, i.e., from the epoch t', determine the credibility of determination of point compatibility.

The numerical size of discrepancies on points or in the geodetic control is assessed on the basis of procedures from the theory of statistical hypothesis verification. Based on these, discrepancies on point are declared either as statistically significant (in the case of higher values of discrepancies) indicating the incompatibility of point, or as statistically insignificant values of random nature, with respect to the methodology of point determination as well as other impacts, testifying to the compatibility of the corresponding point.

All developed statistical verification procedures of compatibility of points contain these principles. The individual methods differ mainly in the use of discrepancy function, its statistical properties, structure and content of testing statistics, type of the probability distribution, etc. (Koch 1988; Kubáček 1978; Kubáčková 1984; Hald 1972).

The majority of testing procedures have a common theoretical background: principles and procedures for the verification of linear hypotheses about the parameters in linear models of the estimation theory (Koch 1988; Kubáčková 1984; Niemeier 1980 and others).

In the following sections, principles and dominant characteristics of some well-known and approved verification procedures that are appropriate for the analysis of the compatibility of geodetic controls and whose detailed interpretation is contained in the relevant cited literature will be listed to provide a brief overview of established methods and procedures.

Several of them are suitable for correct solutions (especially in terms of access to pre-information), others represent solutions with various elements of approximation, and some have advantages mainly from the theoretical point of view and with some of them, sufficient practical experience does not yet exist.

For example, Just (1979) proposes a method for verification of 2D geodetic controls, i.e., the verification of known coordinates of their points only on the basis of their current survey and adjustment. The accuracy of the "old" (epoch t) geodetic control is characterized only by a certain "suitable" and "artificially" formed covariance matrix, established at the epoch t'. The compatibility of points is assessed on the basis of results of analyzes of their accuracy with the appropriate use of S-transformations (also in conjunction with an internal and external reliability of geodetic control) and testing of its mathematical model. As a result of using approximate values, the procedure is not always correct.

The similar approximate solution is proposed by Bill (1984). In the verification of compatibility of p points P from the superior geodetic control with known coordinates $\mathbf{C}$, after the new determination of their coordinates $\hat{\mathbf{C}}'$ (epoch t'), the following localization statistics is used to test coordinate discrepancies $d\mathbf{C} = \hat{\mathbf{C}}' - \mathbf{C}$:

$$T_i = \frac{[\,V_{X_i} \quad V_{Y_i}\,]\,\mathbf{Q}_{\hat{\mathbf{C}}'_i}^{-1}\,[\,V_{X_i} \quad V_{Y_i}\,]^{\mathrm{T}}}{2\bar{s}_0^2} \sim F(2, \infty), \tag{6.5}$$

where the approximate numerical estimation $\bar{s}_0^2$ is used for the unknown variance s_0^2, usually on the basis of present knowledge and experience with the use of certain superior geodetic control. This substitution of s_0^2 for $\bar{s}_0^2$, or substitution of $\mathbf{Q}_{\hat{\mathbf{C}}'_i}$ for $\mathbf{Q}_{\mathbf{C}_i}$, represents the considerable degree of approximation in the verification of points P.

In the situation with full pre-information, the statistics have the following form:

$$T_i = \frac{[\,V_{X_i} \quad V_{Y_i}\,]\,\mathbf{Q}_{\hat{\mathbf{C}}'_i}^{-1}\,[\,V_{X_i} \quad V_{Y_i}\,]^{\mathrm{T}}}{2s_0^2} \sim F(2, n - u), \tag{6.6}$$

by using which, coordinate discrepancies $V_{X_i} = \hat{X}'_i - X_i$, $V_{Y_i} = \hat{Y}'_i - Y_i$ on points P can be correctly assessed.

The solution provided by Koch (1975a, b) gives a specific testing procedure based on the adjusted general linear hypothesis, with differences between "non-random" ("given") coordinates $\mathbf{C}$ at the epoch t and "random" coordinates $\mathbf{C}'$ from the current survey at the epoch t'. The coordinate credibility of point i expressed by the null hypothesis:

$$H_{0i} = \mathbf{E} \begin{bmatrix} V_x = X - X' \\ V_y = Y - Y' \end{bmatrix} = 0 \tag{6.7}$$

is verified by localization statistics:

$$T_i = \frac{(C - C')_i^{\mathrm{T}} \mathbf{Q}_{C'_i}^{-1} (C - C')_i}{2s_0^2} \sim F(2, n - u), \tag{6.8}$$

where s_0^2 is the a posteriori variance factor and

$$\mathbf{Q}_{C'_i} = \begin{bmatrix} q_{X'X'} & q_{X'Y'} \\ q_{Y'X'} & q_{Y'Y'} \end{bmatrix} \tag{6.9}$$

is the cofactor matrix of coordinates C'_i, n is the number of measurements, and u is the number of determined parameters at the epoch t'.

Hanke (1988) proposes an examination of compatibility of p points P determined at the epoch t with coordinates $\mathbf{C}$ so that the current accurate survey of points and determination of $\mathbf{C}'$ (epoch t'), their affine transformation to $\mathbf{C}^t$, determination of discrepancies $\mathbf{V} = [V_X \quad V_Y]$ on points P and their localization testing with the following statistics is carried out:

$$T_i = \frac{V_i^2}{2s_{V_i}^2} \sim F(2, 2p - k - r), \quad k = 6, r = 2, \tag{6.10}$$

where

$$s_{V_i}^2 = \frac{\mathbf{V}^T\mathbf{V} - \frac{V_{X_i}^2 + V_{Y_i}^2}{q_{V_i}}}{2p - k - r} \cdot q_{V_i} \tag{6.11}$$

and with the critical value:

$$F_\alpha = F(1 - \alpha; r, 2p - k - r). \tag{6.12}$$

According to the proposal of Charamza (1975), a defective point in the geodetic control of p points is identified on the basis of coordinate discrepancies $V_X = X - X'$, $V_Y = Y - Y'$ (X, Y given coordinates, X', Y' current coordinate from Helmert transformation). Simple analytical relations to determine proportions of examined points (identical) in the used quadratic form of coordinate discrepancies are given for identification:

$$\mathbf{R} = \sum_{i=1}^{p} (V_{Xi}^2 + V_{Yi}^2) \tag{6.13}$$

and the point with maximum proportion in $\mathbf{R}$ is declared as a defective point. The solution is with no statistical aspects and assessments.

In the work Fotiou et al. (1993), the authors present an algorithm and testing procedures for two independent sets of 2D coordinates of common points with respect to the two reference coordinate systems when using an affine and similarity transformation, which ensure the connection of both systems. This solution is particularly about taking into account the estimates of various components of s_0^2 variance in the stochastic model of observations.

Among the other methods and procedures for verification of survey control points with exact approaches to the identification of character of points (especially using transformation solutions), also the works by Kubáček (1970, 1996) and Teunissen (1986), procedures specified for 2D and 3D points by Koch (1985) with a completion by Benning (1985) as well as various solutions proposed by Gründig (1985), Boljen (1986) and Ingeduld (1985, 1988) should be mentioned. Methods of congruency of point structures are amply developed by Niemeier (1980), Pelzer (1971, 1980), Biacs (1989), and Caspary (1987); they are especially suitable for situations with full pre-information. For situations with incomplete pre-information, mainly procedures by Bill (1985) and Lenzmann (1984) that, although having formal differences, are identical in terms of principles of solution and results, are used from the relevant methods. Analyzes, experiences, and various modifications of certain procedures of this issue are also mentioned by, for example, Jindra (1990), Jakub et al. (1999), Sabová et al. (1999), and Šütti (1996, 1999).

Several of the listed and approved procedures for cases with full as well as partial pre-information are presented in the user version also with examples in the relevant chapters.

6.2.2.2   Methods of Verification with Full Pre-information

For network situations of this type, it is possible to use some of the known and approved procedures, of which the method of planimetric (coordinate) congruency for the structure of geodetic control (Pelzer 1971, 1980; Niemeier 1980) that is widespread also in deformation monitoring and analyses (Pelzer 1971; Caspary 1987; Welsch 1983, 2000; Biacs 1989; Antonopoulis 1985; Chrzanowski 1996 and others) is often implemented.

Method of Planimetric Congruency

• In General

The following situation, in which we want to verify the compatibility of points on the basis of their coordinate congruency, is characteristic for a geodetic control with full pre-information.

Let the local network with points 1, 2, 3, 4, 5 (Fig. 6.3) be created (surveyed, adjusted) in the recent epoch t, and let this network be re-surveyed and re-adjusted with positions of points $1', 2', 3', 4', 5'$ at the epoch t' (at present). Situations of points will be different from each other, thus resulting in two geometric structures of the planimetric network. These structures, if changes of points did not occur in the period $t' - t$ and points were determined correctly at both epochs, will be identical, congruent.

However, in real, two structures of planimetric network will never have an absolutely identical geometry (they will not be theoretically congruent), since coordinates of points at both epochs are influenced by errors of measurement and adjustment as well as by their possible physical changes over time $t' - t$, as well as by other circumstances. When indicators $d\mathbf{X} = X' - X, d\mathbf{Y} = Y' - Y$ represent small values of random character, we can accept the practical congruency of both

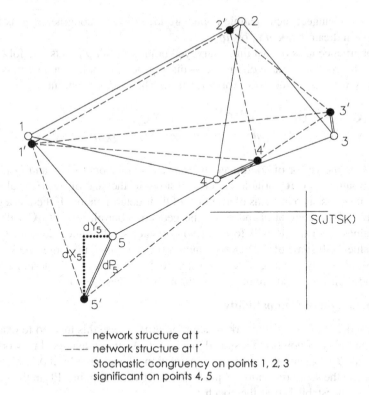

——— network structure at t
——— network structure at t'
Stochastic congruency on points 1, 2, 3
significant on points 4, 5

Fig. 6.3 Planimetric congruency of network structures from epochs *t* and *t'*

realizations of the network, i.e., we are referring to the planimetric stochastic congruency of both structures at *t* and *t'*. However, if dX, dY will represent high values beyond the expectable stochastic degree for some point(s), indicating a planimetric non-identity of point(s), we are referring to the non-congruency of both structures.

In the point set of the network in Fig. 6.3, points 1, 2, and 3 demonstrate acceptable, stochastic congruency, while points 4 and 5 are non-congruent, causing non-congruency of the whole structure.

Congruency of 2D geometry can be verified by appropriate statistical tests that, in terms of compatibility of points at the time $t' - t$, indicate either an acceptable stability, practical identity in the position of points and geometric conformity between survey marks and coordinates, or indicate a situation where at least one point in the structure significantly violated stochastic identity of network geometries by its high values of dX, dY (incompatible point), i.e., it caused their apparent geometric difference—non-congruency.

These facts in the congruency of planimetric networks between different epochs can be used to identify those points that statistically significantly changed their position or were incorrectly determined in the relevant period $t' - t$. These positionally changed, incompatible points are identified by congruence tests using coordinate indicators $d\mathbf{X}, d\mathbf{Y}$, or $d\mathbf{p}$.

In congruence tests of two times surveyed point set with p points, the following quadratic form of coordinate differences—discrepancies of points in the set, is most commonly used as the decision variable on the character of congruency:

$$\mathbf{R} = (\mathbf{C} - \mathbf{C}')^{\mathrm{T}}(\mathbf{Q_C} + \mathbf{Q_{C'}})^{-1}(\mathbf{C} - \mathbf{C}') = d\mathbf{C}^{\mathrm{T}}\mathbf{Q}_{d\mathbf{C}^{-1}}d\mathbf{C}$$
$$\mathbf{C} = [X\ Y], \mathbf{C}' = [X'\ Y'], d\mathbf{C} = [X - X',\ Y - Y'] = [dX\ dY] = [\mathbf{V}_X\ \mathbf{V}_Y] \quad (6.14)$$

where $d\mathbf{C}$ is the vector of coordinate differences—indicators (3.11), and $\mathbf{Q}_{d\mathbf{C}}$ is its cofactor matrix. The $\mathbf{R}$ variable is very sensitive to the size of $d\mathbf{X}, d\mathbf{Y}$ values of individual points, also in terms of accuracy of their determination. If there is a good, acceptable congruency at all points in the geodetic control, i.e., all $d\mathbf{C}$ will have small values, also the $\mathbf{R}$ will form an appropriate, small value. Conversely, the larger values will some of $d\mathbf{C}$ reach (at incompatible points), the larger the value of $\mathbf{R}$ will be. Naturally, the size of $\mathbf{R}$ will be affected by accuracies of determination of $\mathbf{C}, \mathbf{C}'$ and their correlation relationships through matrices $\mathbf{Q_C}, \mathbf{Q_{C'}}$.

- **Examination of Compatibility**

The model situation of network structure in order to use this method to examine the compatibility of points is illustrated in Fig. 6.1. An individual local network LN with points B (e.g., for the needs of construction of various objects), which was connected to the superior (national) planimetric network, to points PL in the system S(JTSK), was established at the epoch t.

At the epoch t' (at present), this LN will be expanded or densified for various reasons (completion, reconstruction, deformation monitoring, etc.). Points B and PL from LN are used as datum (connecting) points for the construction and establishment of new points. Therefore, it is necessary to verify these points in terms of their compatibility, thus to determine whether their physical position has moved at the inter-epoch $t' - t$ and whether they were incorrectly determined at epochs t, t', so that new points in the LN, using PL and B points, can be reliably and accurately determined.

Thus, pre-information and data on LN with points PL, B from the epoch t are available to complete the problem:

$$\mathbf{C}_{\mathrm{PL}}, \mathbf{C}_B, \mathbf{Q}_{\mathrm{C(PL}+B)}, s_0^2, \mathbf{L}, \mathbf{Q_L}, \mathbf{V}, \ldots \quad (6.15)$$

and also data from the realized new survey of the LN' at the epoch t' (measurements "copied" from t):

$$\mathbf{C}'_{\mathrm{PL}}, \mathbf{C}'_B, \mathbf{Q}_{\mathrm{C'(PL}+B)'}, s_0'^2, \mathbf{L}', \mathbf{Q_{L'}}, \mathbf{V}', \ldots \quad (6.16)$$

while also the same datum, i.e., the same connecting points, was used for both epochs t and t'. However, coordinate realizations of the LN network structure (even in the case of identical datum points) will be different from both epochs; they will not be absolutely situationally congruent.

Therefore, the following required variables from the adjustment of network with points B, PL will be used for the application of congruence test from epochs t, t':

$$\hat{\mathbf{C}}, \mathbf{Q}_{\hat{\mathbf{C}}}, s_0^2, \mathbf{L}, \mathbf{Q}_{\mathbf{L}}, \mathbf{V}, n, u, \ldots$$
$$\hat{\mathbf{C}}', \mathbf{Q}_{\hat{\mathbf{C}}'}', s_0'^2, \mathbf{L}', \mathbf{Q}_{\mathbf{L}'}, \mathbf{V}', n', u', \ldots \tag{6.17}$$

having their usual meaning:

$\mathbf{C}, \mathbf{C}'$	coordinate estimates of p points in LN and LN$'$,
$\mathbf{Q}_{\hat{\mathbf{C}}} = (\mathbf{A}^{\mathrm{T}}\mathbf{Q}_{\mathbf{L}}^{-1}\mathbf{A})^{-1}, \quad \mathbf{Q}_{\hat{\mathbf{C}}'}' = (\mathbf{A}'^{\mathrm{T}}\mathbf{Q}_{\mathbf{L}'}^{-1}\mathbf{A}')^{-1}$	cofactor matrices of coordinates, where $\mathbf{A}, \mathbf{A}'$ are configuration matrices of networks LN and LN$'$,
$s_o^2, s_0'^2$	a posteriori unit variances from adjustments of LN, LN$'$,
$\mathbf{L}, \mathbf{L}'$	vectors of measured variables,
$\mathbf{Q}_{\mathbf{L}}, \mathbf{Q}_{\mathbf{L}'}$	cofactor matrix of vectors $\mathbf{L}, \mathbf{L}'$,
$\mathbf{V}, \mathbf{V}'$	residuals (corrections) of measured variables,
n, n'	number of measured variables,
p	number of points in LN and LN$'$,
$u = u' = 2p$	number of determined coordinates of p points, etc.

Differences—coordinate indicators (coordinate discrepancies) are determined from coordinates[1]:

$$\mathbf{V_C} = d\mathbf{C} = \mathbf{C}' - \mathbf{C} = \begin{bmatrix} dC_1 \\ \vdots \\ dC_i \\ \vdots \end{bmatrix} = \begin{bmatrix} X_1' \ Y_1' \\ \vdots \\ X_i' \ Y_i' \\ \vdots \end{bmatrix} - \begin{bmatrix} X_1 \ Y_1 \\ \vdots \\ X_i \ Y_i \\ \vdots \end{bmatrix} = \begin{bmatrix} dX_1 \ dY_1 \\ \vdots \\ dX_i \ dY_i \\ \vdots \end{bmatrix} \tag{6.18}$$

---

[1]Coordinate "corrections"—discrepancies (differences) at points will be referred to as dC or $\mathbf{V_C}$, corrections (residues) of measured values $\mathbf{L}$ as $\mathbf{V}$.

and their cofactor matrix (Koch 1988):

$$
\mathbf{Q}_{dC} = \mathbf{Q}_C + \mathbf{Q}_{C'} =
\begin{bmatrix}
\mathbf{Q}_{dC_{11}} & \cdots & \mathbf{Q}_{dC_{1i}} & \cdots & \mathbf{Q}_{dC_{1p}} \\
\vdots & & \vdots & & \vdots \\
\mathbf{Q}_{dC_{i1}} & \cdots & \mathbf{Q}_{dC_{ii}} & \cdots & \mathbf{Q}_{dC_{ip}} \\
\vdots & & \vdots & & \vdots \\
\mathbf{Q}_{dC_{p1}} & \cdots & \mathbf{Q}_{dC_{pi}} & \cdots & \mathbf{Q}_{dC_{pp}}
\end{bmatrix},
\tag{6.19}
$$

where submatrices are

$$
\mathbf{Q}_{dC_{ii}} =
\begin{bmatrix}
q_{dx_i} & q_{dxy_i} \\
q_{dyx_i} & q_{dy_i}
\end{bmatrix}.
\tag{6.20}
$$

In addition to these, following are determined for testing:

- quadratic forms of residuals—corrections of measured variables $\mathbf{L}, \mathbf{L}'$ (from both epochs):

$$
\begin{aligned}
\Omega &= \mathbf{V}^T \mathbf{Q}_L^{-1} \mathbf{V} \\
\Omega' &= \mathbf{V}'^T \mathbf{Q}_{L'}^{-1} \mathbf{V}',
\end{aligned}
\tag{6.21}
$$

- quadratic forms of coordinate differences: global $\mathbf{R}$ (5.14) and point $\mathbf{R}_i$ (approximate):

$$
\begin{aligned}
\mathbf{R} &= d\mathbf{C}^T \mathbf{Q}_{dC}^{-1} d\mathbf{C} \\
\mathbf{R}_i &= d\mathbf{C}_i^T \mathbf{Q}_{dC_i}^{-1} d\mathbf{C}_i = (C' - C)_i^T \mathbf{Q}_{dC_i}^{-1} (C' - C)_i,
\end{aligned}
\tag{6.22}
$$

- and common a posteriori unit variance from both epochs:

$$
\bar{s}_o^2 = \frac{\Omega + \Omega'}{(n - u) + (n' - u')}.
\tag{6.23}
$$

Testing statistics represents either a random variable (Niemeier 1980; Pelzer 1971):

$$
T = \frac{d\mathbf{C}^T \mathbf{Q}_{dC}^{-1} d\mathbf{C}}{f_G \bar{s}_o^2} = \frac{\mathbf{R}}{f_G \bar{s}_o^2} \sim F(f_G, f_G')
\tag{6.24}
$$

for the so-called global congruence test of network realizations LN and LN' (with Gauss–Markov model with full rank), which has the Fisher–Snedecor distribution

(F-distribution) of probability with degrees of freedom (using the regular Gauss–Markov model of adjustment):

$$f_G = \text{rank}(\mathbf{Q_C} + \mathbf{Q}_{C'}) = u, \ f'_G = (n+n') - 2u \tag{6.25}$$

and either statistics:

$$T_i = \frac{d\mathbf{C}_i^T \mathbf{Q}_{d\mathbf{C}i}^{-1} d\mathbf{C}_i}{f_L \bar{s}_o^2} = \frac{\mathbf{R}_i}{f_L \bar{s}_o^2} \sim F(f_L, f'_L), \tag{6.26}$$

of the so-called localization (point) test for individual points in the network realization, which also has the F-distribution with degrees of freedom:

$$f_L = 2, f_{L'} = f_{G'} = (n+n') - 2u. \tag{6.27}$$

A global test with the statistics (6.24) is used as the first step of testing procedure for the overall assessment of compatibility of all points in the LN, whose possible states are expressed in the form of null and alternative hypothesis:

$$H_0 : d\mathbf{C} = 0, H_a : d\mathbf{C} \neq \cong 0. \tag{6.28}$$

H_0 declares a situation in which compatibility indicators $d\mathbf{C} = \mathbf{C}' - \mathbf{C}$ are practically insignificant, stochastic values for all points in the LN, i.e., the state that there were no significant changes on points in the LN in the period $t' - t$, for any reasons. This implies that all points PL and B in the LN determined at t can be used for geodetic activities also at the epoch t'.

H_a indicates the opposite situation that some of the points B and PL will show evident changes in the period $t' - t$, i.e., they will show up as incompatible.

The following level of significance is chosen for testing:

$$\alpha = 0.05 \text{ or } 0.02, 0.01, 0.005, \ldots, \tag{6.29}$$

the critical value $\eta\{(1 - \alpha)\text{-quantile}\}$ of the F-distribution with degrees of freedom f_G, f'_G is determined:

$$F_{\alpha_G} = F\left(\alpha; F_G, F'_G\right), \tag{6.30}$$

and values of T and F_{α_G} are compared:

if $T < F_{\alpha_G}$,

H_0 is not rejected, a significant change of network congruency, i.e., the incompatibility of its points by their determination at the time t' against the determination at the time t, is not provable. Therefore, points PL and B are considered as practically unchanged in the period $t' - t$, i.e., they are considered as points compatible and suitable for use in the current geodetic activities in the area of LN;

if $T \geq F_{\alpha_G}$,

H_0 is rejected (with the risk α of an incorrect rejection of H_0) and obviously H_a is true; therefore, we can practically assume that there were significant defects and emergence of incompatibility of some point(s) PL, B was also reflected in their coordinates, in the period between epochs t and t' in the physical geodetic control of the LN network (or even during its survey at the epoch t).

In this case, we can proceed to the localization test, i.e., every point is verified within its algorithm, whether it has a role in the establishment of non-congruency of network structure (establishment of incompatibility of geodetic control). Thus, in this testing procedure, point(s) that is incompatible and caused the rejection of H_0 for the whole geodetic control by the global test (6.24) is identified by T_i statistics (6.26).

Localization testing is realized using a point quadratic form of coordinate differences (discrepancies) $\mathbf{R}_i$ (6.22), which is used in the statistics (6.26) and which represents proportions of individual points of LN on the total values of $\mathbf{R}$. The larger this proportion will be, the more likely the ith point will have non-congruent, i.e., incompatible character.

However, instead of the above procedure of local testing, a more time-saving method is used most commonly, for example, the localization testing with an algorithm of the global test in cycles, while eliminating the maximum values of $\mathbf{R}_i$ from $\mathbf{R}$ (in detail in Sect. 7.3.1).

- **Determination of $\mathbf{R}_i$**

Values of $\mathbf{R}_i$ for a certain geodetic control can be determined at two levels: approximately and exactly.

In most cases, even approximate values of $\mathbf{R}_i$ are suitable for common practice and they usually provide findings identical to results and findings of exact testing, regarding the compatibility of points, in testing procedures.

When using the approximate procedure, individual $\mathbf{R}_i$ are obtained according to (6.22), and the following relation applies:

$$\mathbf{R} = d\mathbf{C}^{\mathrm{T}}\mathbf{Q}_{d\mathbf{C}}^{-1}d\mathbf{C} \cong\neq \Sigma\, d\mathbf{C}_i^{\mathrm{T}}\mathbf{Q}_{dC_i}^{-1}d\mathbf{C}_i,\ i = 1, 2, \ldots, p, \tag{6.31}$$

which, based on the principle of their determination, do not take into account the effects of off-diagonal cofactors from the $\mathbf{Q}_{dC}$ matrix, i.e., they do not take into account auto- and inter-correlation relations between coordinates of points and their effects.

Exact values of $\bar{\mathbf{R}}_i$ are determined on the basis of decomposition of the quadratic form of coordinate indicators $\mathbf{R}$ (Pelzer 1971; Caspary 1987; Biacs 1989), according to which, the value of $\bar{\mathbf{R}}_i$ is allocated from $\mathbf{R}$ for the ith point according to:

$$\mathbf{R} = \mathbf{R}_n + \bar{\mathbf{R}}_i, \tag{6.32}$$

where $\mathbf{R}_n$ is the proportion of other points in $\mathbf{R}$. Further,

$$\bar{\mathbf{R}}_i = d\bar{\mathbf{C}}_i^{\mathrm{T}} \mathbf{Q}_{d\mathbf{C}_i}^{-1} d\bar{\mathbf{C}}_i, \tag{6.33}$$

where modified coordinate indicators are

$$d\bar{\mathbf{C}}_i = \begin{bmatrix} d\bar{X} \\ d\bar{Y} \end{bmatrix}_i = d\mathbf{C}_i + \mathbf{Q}_{d\mathbf{C}_{ii}} \mathbf{Q}_{d\mathbf{C}_{in}}^{-1} d\mathbf{C}_n, \tag{6.34}$$

while matrices $\mathbf{Q}_{d\mathbf{C}_{ii}}, \mathbf{Q}_{d\mathbf{C}_{in}}$ result from the adjustment of the matrix:

$$\mathbf{Q}_{d\mathbf{C}}^{-1} = \begin{bmatrix} \mathbf{Q}_{d\mathbf{C}_{nn}}^{-1} & \mathbf{Q}_{d\mathbf{C}_{ni}}^{-1} \\ \mathbf{Q}_{d\mathbf{C}_{in}}^{-1} & \mathbf{Q}_{d\mathbf{C}_{ii}}^{-1} \end{bmatrix} \tag{6.35}$$

within the cyclic change of points with $d\mathbf{C}_i$ in the vector $d\mathbf{C}$. According to thus obtained relations, $\bar{\mathbf{R}}_i$ take into account all the necessary correlation bonds from the matrix $\mathbf{Q}_{d\mathbf{C}}$. By this algorithm, variables (6.33) and (6.34) necessary for testing are determined by the outlined procedure from the vector $d\mathbf{C}$ for each element of $d\mathbf{C}_i$ and point.

However, even the approximate procedure to determine $\mathbf{R}_i$ according to (6.22) conforms well to the objectives, i.e., to the identification of incompatibility of points. The exact solution with $\bar{\mathbf{R}}_i$ is especially desirable in the analysis of deformation monitoring (Pelzer 1980; Caspary 1987; Welsch 2000; Chrzanowski 1996).

- **Localization of Incompatible Point**

The algorithm for the localization of incompatible point thus involves successive testing of every point in order to identify that point, whose incompatibility caused the rejection of H_0 in the global test (6.24).

The null hypothesis H_{0i} (and the relevant alternative hypothesis H_{ai}) is also accepted for the localization test:

$$H_{0i} : d\mathbf{C}_i = 0, H_{ai} : d\mathbf{C}_i \neq 0, \tag{6.36}$$

which postulates that the coordinate indicator of a point from the network structure of LN indicates the compatibility of point based on its insignificant, stochastic value. This hypothesis is tested on the level of significance (Bill 1985; Koch 1975):

$$\alpha_0 = 1 - (1 - \alpha)^{1/P} \cong \frac{\alpha}{p} \tag{6.37}$$

by using the statistics (6.26), whose value is quantified by using the components (6.18), (6.19), (6.22), (6.23). The critical value $\{(1 - \alpha_0)$-quantile of the F-distribution$\}$ for degrees of freedom f_L, f_L' corresponds to this hypothesis:

$$F_{\alpha_i} = F\left(\alpha_0; f_L, f_L'\right). \tag{6.38}$$

However, $\alpha_0 = 0,01$ is most commonly used regardless of p.

If $T_i < F_{\alpha_i}$,

H_{0i} is not rejected and we accept that the tested point B is compatible with the epoch t';

if $T_i \geq \Sigma F_{\alpha_i}$,

H_{0i} is rejected and we consider the examined point PL or B as the cause of the significant defect of positional congruency of network LN, i.e., we consider the relevant point as incompatible with the structure of LN, and it cannot be used for geodetic activities in the area.

Example No. 1 An independent local network LN with 5 points B and 3 datum points PL (Fig. 6.4) was established and surveyed for the purpose of surveying securing of industrial buildings construction at the epoch t. The network was adjusted by the connection with Gauss–Markov model with full rank (regular Gauss–Markov model) to 3 surrounding datum (connecting) points PL of the National Triangulation Network—ŠTS with coordinates in the S(JTSK).

Coordinates $\mathbf{C}$ of determined points B at the epoch t are shown in Table 6.1: and necessary additional data from the survey and adjustment (6.17):

$$\Omega = 0.008463785,$$
$$n = 18,$$
$$u = 10,$$

as well as 10×10 cofactor matrix of coordinate estimates according to $\mathbf{Q}_{\hat{C}} = \left(\mathbf{A}^T \mathbf{Q}_L^{-1} \mathbf{A}\right)^{-1}$ (not specified).

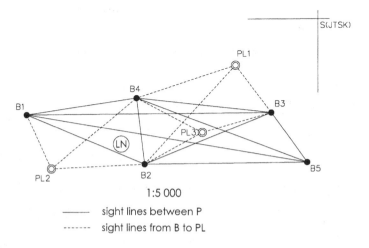

Fig. 6.4 The local network LN of points B and PL determined at t and t'

At the epoch t', points B should be used to establish net densification points in the structure of LN and it is necessary to determine by its actual survey whether some points B of the network LN degraded during the period $t' - t$, whether they are also compatible at the epoch t' and thus whether it will be possible to connect new, determined points to points B of the network LN from the epoch t at the epoch t'.

Coordinates $\mathbf{C}'$ of determined points at the epoch t' are shown in Table 6.2: and necessary additional data from the survey and adjustment (6.17):

$$\Omega' = 0.004823507,$$
$$n' = 19,$$
$$u' = 10,$$

as well as 10×10 cofactor matrix of coordinate estimates is known from the adjustment according to $\mathbf{Q}_{\hat{C}'} = \left(\mathbf{A}'^{T}\mathbf{Q}_{L'}^{-1}\mathbf{A}'\right)^{-1}$ (not specified).

Coordinate indicators (discrepancies) $\mathbf{V_C} = d\mathbf{C} = \mathbf{C}' - \mathbf{C}$ from values in Tables 6.1 and 6.2 are shown in Table 6.3: their cofactor matrix $\mathbf{Q}_{d\mathbf{C}}$ (not specified) was determined according to (6.19), the common a posteriori unit variance according to (6.23) with the value of:

$$\bar{s}_0^2 = \frac{0.008463785 + 0.004823507}{(18 - 10) + (19 - 10)} = 0.0007816054\ 12,$$

and quadratic forms of coordinate, global and point indicators according to (6.22) with values of:

$$\mathbf{R} = 0.0159276,$$

and (Table 6.4):

Table 6.1 Coordinates of determined points

Point	X (m)	Y (m)
B1	1,205,419.082	282,561.296
B2	1,205,505.831	282,352.152
B3	1,205,416.008	282,127.888
B4	1,205,390.044	282,366.089
B5	1,205,503.033	282,064.664

Table 6.2 Coordinates of points determined at subsequent epoch

Point	X' (m)	Y' (m)
B1	1,205,419.073	282,561.309
B2	1,205,505.838	282,352.145
B3	1,205,416.001	282,127.898
B4	1,205,390.052	282,366.081
B5	1,205,503.036	282,064.658

Table 6.3 Coordinate indicators—discrepancies

Point	dX (m)	dY (m)
B1	−0.009	0.013
B2	0.007	−0.007
B3	−0.007	0.010
B4	0.008	−0.008
B5	0.003	−0.006

Table 6.4 Values of R at individual points

Point	$\mathbf{R}_i$
B1	0.00320347
B2	0.00526195
B3	0.00128771
B4	0.00246520

The test statistic for the global test according to (6.24) will be

$$f_G = u = 10, f_G' = (18 + 19) - 2 \times 10 = 17,$$

$$T = \frac{\mathbf{R}}{f_G \bar{s}_0^2} = \frac{0.0159276}{10 \times 0.000781605412} = 20{,}378,$$

and its critical value according to (6.30) for the level of significance $\alpha = 0.05$:

$$F_{\alpha_G} = F(0.05; 10.17) = 2.4499.$$

The comparison of T and F_{α_G} gives:

$$T < F_{\alpha_G},$$

therefore, H_0 is not rejected with the relevant statistical interpretation of this inequality, all 5 points determined from two independent measurements at the epoch t and once again surveyed and determined for verification at the epoch t' represent compatible points (points with no significant displacements or their incorrect determination) with the possibility of their use for actual geodetic activities in that area.

Results of localization tests for individual points (although localization testing is no longer required for the result $T < F_{\alpha_G}$) realized according to (6.26) with variables (6.27), (6.22), (6.23) also indicate the compatibility of 5 verified points. For point test results, we get:

$$f_L = 2, f_L' = f_G' = (18 + 19) - 2 \times 10 = 17, \alpha_0 = 0.05/5 = 0.01,$$

and

$$T_i = \frac{\mathbf{R}_i}{f_L \bar{s}_0^2} = \frac{\mathbf{R}_i}{2 \times 0.000781605412} = \begin{bmatrix} 2.0493 \\ 3.3661 \\ 0.8238 \\ 1.5770 \\ 4.0497 \end{bmatrix}.$$

For critical value of the test statistic (6.38) with the use of approximate values of $\mathbf{R}_i$ according to (6.31):

$$F_{\alpha_i} = F(0.01; 2.17) = 6.1121$$

and therefore,

$$\forall T_i < F_{\alpha_i},$$

i.e., all points B prove their compatibility also on the basis of localization testing using approximate values of $\mathbf{R}_i$.

Boljen's Method

In cases where the situation is analogous as described in the previous Sect. Method of Planimetric Congruency with the exception that at this time, different datum points were used at epochs t and t', i.e., network structures at t and t' have different datum realization of reference coordinate frame, it is appropriate to use other methods, for example, Boljen's method (1986), to assess the compatibility of points PL, B. By this method, it can be reliably determined at which points in the LN there is no stochastic consistency between physical survey marks and coordinates due to displacement of point and incorrect determination of coordinates.

In the realization of this method, as with the congruent method, all required variables, and data (6.17) are determined by conventional adjustment procedures. However, coordinate differences $d\mathbf{C} = \hat{\mathbf{C}}' - \hat{\mathbf{C}}$ on points PL, B are not directly used to assess their compatibility, because they are also influenced by different datums of $\hat{\mathbf{C}}$ and $\hat{\mathbf{C}}'$ determination. This datum influence can be (and it will always be necessary) eliminated (or significantly suppressed) by transformation of coordinates $\hat{\mathbf{C}}'$ [from the realization of coordinate frame $S'(X'Y')$] to coordinates $\mathbf{C}^t = [X^tY^t]$ [to the coordinate frame $S(XY)$]. A standard Helmert transformation (with supernumerary identical points) or even its differential version (Mierlo 1980; Illner 1983) is commonly used for such transformation operation.

Coordinate discrepancies resulting from transformation

$$\mathbf{V}_{\mathbf{C}_i} = d\mathbf{C}_i = \hat{\mathbf{C}}_i - \mathbf{C}_i^t = \begin{bmatrix} \mathbf{V}_X \\ \mathbf{V}_Y \end{bmatrix}_i = \begin{bmatrix} \hat{X} - X^t \\ \hat{Y} - Y^t \end{bmatrix}_i, \quad i = 1, 2, \ldots, p \qquad (6.39)$$

create the subject of testing, since it is a natural, logical postulate that large values of $\mathbf{V}_{\mathbf{C}} = [\mathbf{V}_X \mathbf{V}_Y]$ will identify point with significantly violated compatibility, while small values of $\mathbf{V}_{\mathbf{C}}$ (of stochastic size) will represent a natural attribute of processes of point determination.

The testing procedure is realized in cycles with the global test (6.41), in which it is found, whether also incompatible points can be found in the whole geodetic control of points PL, B. These incompatible points are identified by the localization test (6.43).

The null and alternative hypothesis is formulated for the realization of global test:

$$H_0 : \mathbf{V}_{\mathbf{C}} = 0; H_a : \mathbf{V}_{\mathbf{C}} \neq 0 \qquad (6.40)$$

and test statistic that has the F-distribution (Boljen 1986) is used:

$$T = \frac{6\mathbf{V}_{\mathbf{C}}^T\mathbf{V}_{\mathbf{C}}}{\sqrt{p}(2p - \text{tp})(s_0^2 + m^2 s_0'^2)} \sim F(f_1, f_2), \qquad (6.41)$$

where

$\mathbf{V}_{\mathbf{C}} = \begin{bmatrix} V_{X1} & V_{Y1} & \ldots & V_{Xp} & V_{Yp} \end{bmatrix}^T$	are coordinate discrepancies on individual identical points (from transformation),
$s_0^2 = \frac{\mathbf{V}^T\mathbf{Q}_{\mathbf{L}}^{-1}\mathbf{V}}{n-u}$	is the a posteriori unit variance from the adjustment of local network LN at the epoch t,
$s_0'^2 = \frac{\mathbf{V}'^T\mathbf{Q}_{\mathbf{L}'}^{-1}\mathbf{V}'}{n'-u'}$	is the a posteriori unit variance from the adjustment of local network LN at the epoch t', in which $\mathbf{V}$ is the vector of corrections (residuals) of observations $\mathbf{L}$ at the epoch t and $\mathbf{V}'$ for $\mathbf{L}'$ at the epoch t',
$M = 1$	is a scale denominator,
p	is the number of points B at t and t',
n, n'	is the number of measurements at t and t',
$u = u' = 2p$	is the number of coordinates of p determined points B,
tp	is the number of transformation parameters,
$f_1 = 2p - \text{tp}, f_2 = \infty$	are degrees of freedom of the F-distribution

Critical value of the F-distribution for degrees of freedom f_1, f_2 is determined by the choice of level of significance of testing α, usually with values (6.29):

$$F_\alpha = F(1 - \alpha; f_1, f_2). \qquad (6.42)$$

In the **0th cycle**, if the following results from the comparison (6.41), (6.42):

$$T < F_\alpha,$$

H_0 is not rejected and, therefore, the whole set of points PL and B in the LN will be considered as compatible points, i.e., points with only stochastically changed coordinates during the period $t' - t$ and thus eligible for their use in the current geodetic activities;

if $T \geq F_\alpha$,

H_0 is rejected (with the risk α of incorrect rejection of H_0); therefore, points in the LN cannot be considered as compatible because an incompatible point (or points) that has to be identified will be among them.

The localization of an incompatible point is realized so that the corresponding point value T_i related to one, ith point (Boljen 1986) is determined for each identical point:

$$T_i = \frac{s_0^2(t)}{\left(\frac{\sqrt{p}}{6}\right)(s_0^2 + m^2 s_0^2)} \cdot \frac{\mathbf{V}_{\mathbf{C}_i}^{\mathrm{T}} \mathbf{V}_{\mathbf{C}_i}}{\frac{p-1}{p} - \frac{X_i'^2 + Y_i'^2}{\sum_1^p (X_i'^2 + Y_i'^2)}}, \qquad (6.43)$$

where $s_0^2(t)$ is the a posteriori unit variance from transformation, discrepancies $\mathbf{V}_{\mathbf{C}_i}$ are defined according to (6.39) and X_i', Y_i' are reduced coordinates of identical points $i = 1, 2, \ldots, p$. A postulate is adopted that of determined values $T_i, i = 1, 2, \ldots, p$ of individual points PL, B, always the point with the maximum value of T_i causes the incompatibility of geodetic control, so the relevant point will be considered as incompatible and inapplicable. This point is excluded from the set of p points and $p - 1$ points remain to continue the testing process.

In the **1st cycle**, the remaining $p - 1$ points are examined for compatibility of their geodetic control by the global test (6.41), while the relevant changes caused by deletion of point are reflected in its statistics.

For this purpose, a new transformation of the $p - 1$ remaining points PL, B is performed, and their coordinates $\mathbf{C}^t(1)^2$ and discrepancies $\mathbf{V}_{\mathbf{C}}(1) = \mathbf{C}(1) - \mathbf{C}^t(1)$ as well as the corresponding test statistic $T(1)$ and its critical value $F_\alpha(1)$ are determined.

If $T(1) < F_\alpha(1)$, the remaining points of the geodetic control in the number of $p - 1$ will be compatible.

---

[2]Denotations (2), (3),...,(j) represent corresponding cycles.

If $T(1) \geq F_\alpha(1)$, it is necessary to exclude the point with the maximum value of T_i from the set of $p - 1$ points as another incompatible point by the procedure outlined.

Subsequently, we proceed by specified algorithms in subsequent cycles until the jth cycle, in which the testing will indicate $T(j) < F_\alpha(j)$, i.e., the compatibility of all remaining points.

Example No. 2 Consider a point situation presented in the Example No. 1, so an individual local network in the system S-JTSK with $p = 5$ points B (Fig. 6.4), for which we have adjusted coordinates in Tables 6.1 and 6.2 from measurements at epochs t and t'. Quadratic forms of corrections and a posteriori unit variances from adjustment at epochs t and t' are

$$\Omega = 0.008463785, \quad s_0^2 = \Omega/(n - u) = 0.001057973,$$
$$\Omega' = 0.004823507, \quad s_0'^2 = \Omega'/(n' - u') = 0.000535945.$$

Using all 5 points, tp = 4 transformation parameters for Helmert transformation of coordinates $\hat{\mathbf{C}}'$ into coordinates $\mathbf{C}^t$ were determined, for which the following values were obtained (Table 6.5):and coordinate discrepancies were determined according to $\mathbf{V_C} = \mathbf{C} - \mathbf{C}^t$ using coordinates $\mathbf{C}$ in Table 6.1 (Table 6.6):

For global test according to (6.41), the realization of test statistic T was quantified for $m = 1$ and other known variables with the value of

$$T = 0.3821,$$

the level of significance of the test $\alpha = 0.05$ was selected, and the critical value F_α for degrees of freedom $f_1 = 2 \times 5 - 4 = 6, f_2 = \infty$ according to (6.42) is:

$$F_\alpha = 2.0992.$$

Table 6.5 Coordinates of determined points

Point	X^t (m)	Y^t (m)
B1	1205419.090	282561.291
B2	1205505.827	282352.157
B3	1205416.011	282127.879
B4	1205390.038	282366.093
B5	1205503.035	282064.672

Table 6.6 Coordinate discrepancies of individual points

Point	V_X (mm)	V_Y (mm)
B1	−8	5
B2	4	−5
B3	−3	9
B4	6	−4
B5	−2	−8

Based on obtained values,

$$T < F_\alpha,$$

thus H_0 is not rejected, the whole geodetic control of 5 points B established at the epoch t can be considered as compatible and eligible for the use for further geodetic activities at the current epoch t'.

6.2.2.3 Methods of Verification with Incomplete Pre-information

Cases of verifying the compatibility of existing survey control points in various geodetic controls, where only coordinates of points in a certain system $S(XY)$ from the epoch t are available, create the most difficult cases of examination of the quality of points, i.e., their compatibility. It is also evident that the uncertainty of solution of these situations in terms of objectives of the solution allows us to select and use multiple methods to examine the compatibility of geodetic control or its parts, which may also have a different degree of incorrectness, an approximation. In the specific situation of this type [the pre-information situation (b)], it is, therefore, necessary to use only those methods to obtain a trustworthy image of compatibility of points of the existing methods, which will allow us to create the most realistic view of the pursued objectives by their principles.

Some of these methods, which acquitted well based on their characteristics and provide reliable results, will be presented in next sections for the verification of compatibility of 2D geodetic controls. Deciding on the compatibility or incompatibility of points will require a preparation of all necessary variables and parameters as well as the use of appropriate tests from the group of so-called tests for outlying values of coordinates of examined points (Koch 1983; Kok 1984; Pope 1976; Baarda 1968; Lenzmann 1984; Heck 1985 and others).

Lenzmann–Heck's Method

• The Principle and its Applicability

This method can be suitably used for certain existing geodetic controls with points P (of a superior network), of which, for example, the following tasks should be realized in the area of LN (Fig. 6.2):

(a) points labeled as PL should be verified for compatibility since they will be used for current, planned geodetic activities in the area at the epoch t',

(b) a superior geodetic control should be densified in the area of LN, i.e., in addition to the existing points PL also new points U are established, with the result that the compatibility of PL points is verified (whether these points are eligible as datum—connecting points to determine the set of new points U), as well as other similar tasks and objectives.

A procedure for solving this task, i.e., the verification of compatibility of points PL, is specified for a more general case (b).

Only coordinates $\mathbf{C}_{PL}^{J} = [XY]_{PL}^{J}$ in the system S(JTSK) are available for points P of the superior geodetic control from the epoch t (establishment of the geodetic control, its completion, etc.).

The number u of new points U and the number p of points PL is connected (surveyed) by a suitable network structure on the required level of accuracy in the area of LN at the epoch t'. For the processing and adjustment of measurements, the implementer establish a local coordinate system S(LOC) (e.g., according to Fig. 5.2), basically only by an accurate measurement of the length d(LOC), in which the local network connecting points U and PL is adjusted using the Gauss–Markov regular or singular model, thereby obtaining coordinates $\hat{\mathbf{C}}_{PL}^{L} = [\hat{X}\hat{Y}]_{PL}^{L}$, $\hat{\mathbf{C}}_{U}^{L} = [\hat{X}\hat{Y}]_{U}^{L}$ and other necessary parameters and data for these points.

Points PL will, therefore, have coordinates $\mathbf{C}_{PL}^{J}$ defined in the system S(JTSK) from the epoch t and coordinates $\hat{\mathbf{C}}_{PL}^{L}$ defined in the system S(LOC) at the epoch t'. Connecting points PL are also identical points for the determination of transformation parameters $(\hat{\mathbf{T}}\mathbf{P})$ for the transformation of coordinates $\hat{\mathbf{C}}_{PL}^{L}$ and $\hat{\mathbf{C}}_{U}^{L}$ to corresponding coordinates $\mathbf{C}_{PL}^{Jt} = [XY]_{PL}^{Jt}$ and $\mathbf{C}_{U}^{Jt} = [\hat{X}\hat{Y}]_{U}^{Jt}$ in the system S(JTSK). Subsequently, it is possible to assess differences—coordinate discrepancies $d\mathbf{C}_{PL} = \mathbf{C}_{PL}^{J} - \mathbf{C}_{PL}^{Jt}$ (coordinate indicators) using an appropriate statistical testing and determine which points PL are compatible and which are not, based on coordinates $\mathbf{C}_{PL}^{J}$ and $\mathbf{C}_{PL}^{Jt}$, i.e., coordinates $\mathbf{C}_{PL}^{J}$ of points PL from their initial determination at the epoch t and coordinates $\mathbf{C}_{PL}^{Jt}$ from the current survey and measurement at the epoch t'.

- **Transformation Determination of Data for Testing**

For the outlined procedure, a similarity (Helmert) transformation with the LSM determination (Gauss–Markov regular model) of transformation parameters according to known relationships (Benning 1985; Heck 1985; Šütti et al. 1997; Jakub 2001; Schuh 1987) is most commonly used, in which these parameters are dominant:

– a matrix of coefficients (coordinates $\hat{\mathbf{C}}_{PL}^{L}$ should be reduced to their center of gravity):

$$\mathbf{A}_{PL} = \begin{bmatrix} 1 & 0 & \hat{X}_{PL_1}^{L} & -\hat{Y}_{PL_1}^{L} \\ 0 & 1 & \hat{Y}_{PL_1}^{L} & \hat{X}_{PL_1}^{L} \\ \vdots & & & \\ 1 & 0 & \hat{X}_{PL_p}^{L} & -\hat{Y}_{PL_p}^{L} \\ 0 & 1 & \hat{Y}_{PL_p}^{L} & \hat{X}_{PL_p}^{L} \end{bmatrix}, \tag{6.44}$$

– coordinates of points PL in the S(JTSK) at the epoch t:

$$\mathbf{C}_{PL}^{J} = \begin{bmatrix} X_{PL_1}^{J} \\ Y_{PL_1}^{J} \\ \vdots \\ X_{PL_p}^{J} \\ Y_{PL_p}^{J} \end{bmatrix}. \tag{6.45}$$

By using these, transformation parameters $\hat{\mathbf{TP}}$ are determined according to

$$\hat{\mathbf{TP}} = (\mathbf{A}_{PL}^{\mathrm{T}}\mathbf{A}_{PL})^{-1}\mathbf{A}_{PL}^{\mathrm{T}} \cdot \mathbf{C}_{PL}^{J}, \tag{6.46}$$

and by transformation

$$\mathbf{C}_{PL}^{Jt} = \begin{bmatrix} X_{PL_1}^{Jt} \\ Y_{PL_1}^{Jt} \\ \vdots \\ X_{PL_p}^{Jt} \\ Y_{PL_p}^{Jt} \end{bmatrix} = \mathbf{A}_{PL}\hat{\mathbf{TP}}, \tag{6.47}$$

coordinates $\mathbf{C}_{PL}^{Jt}$ of points PL in the S(JTSK) are obtained.

Based on the transformation, we get

– cofactor matrix of transformation parameters:

$$\mathbf{Q}_{\hat{\mathbf{TP}}} = (\mathbf{A}_{PL}^{\mathrm{T}}\mathbf{A}_{PL})^{-1} = \mathbf{N}^{-1}, \tag{6.48}$$

– coordinate discrepancies (coordinate indicators) $-\mathbf{V}_{\mathbf{C}}$ between coordinates $\mathbf{C}_{PL}^{J}$ and $\mathbf{C}_{PL}^{Jt}$ of datum ($\because$ identical) points PL:

$$\mathbf{V}_{\mathbf{C}} = \mathbf{C}_{PL}^{J} - \mathbf{C}_{PL}^{Jt} = \begin{bmatrix} X_{PL_1}^{J} - X_{PL_1}^{Jt} \\ Y_{PL_1}^{J} - Y_{PL_1}^{Jt} \\ \vdots \\ X_{PL_p}^{J} - X_{PL_p}^{Jt} \\ Y_{PL_p}^{J} - Y_{PL_p}^{Jt} \end{bmatrix} = \begin{bmatrix} V_{X_{PL1}} \\ V_{Y_{PL1}} \\ \vdots \\ V_{X_{PLp}} \\ V_{Y_{PLp}} \end{bmatrix} = \mathbf{C}_{PL}^{J} - \mathbf{A}_{PL}\hat{\mathbf{TP}}, \quad (6.49)$$

– a posteriori unit variance:

$$s_0^2 = \frac{(\mathbf{V}_{\mathbf{C}}^{\mathrm{T}}\mathbf{V}_{\mathbf{C}})}{(2p - 4)}, \tag{6.50}$$

– the covariance matrix of transformation parameters, which provides standard deviations of LSM by estimation of determined transformation parameters:

$$\sum\nolimits_{\hat{\mathbf{T}}\mathbf{P}} = s_0^2 \mathbf{Q}_{\hat{\mathbf{T}}\mathbf{P}} = s_0^2 (\mathbf{A}_{PL}^{\mathrm{T}} \mathbf{A}_{PL})^{-1}. \tag{6.51}$$

With determined transformation parameters tp = 4, also coordinates $\mathbf{C}_U^L = [X\ Y]_U^L$ of newly determined points U in the area of LN at the epoch t' are transformed into coordinates $\mathbf{C}_U^{Jt}$ according to

$$\mathbf{C}_U^{Jt} = \mathbf{A}_U \hat{\mathbf{T}}\mathbf{P} = \begin{bmatrix} X_{U_1}^{Jt} \\ Y_{U_1}^{Jt} \\ \vdots \\ X_{U_u}^{Jt} \\ Y_{U_u}^{Jt} \end{bmatrix}. \tag{6.52}$$

where $\mathbf{A}_U$ is a matrix of coefficients of the analogous structure as (6.44) with elements—reduced coordinates $\hat{\mathbf{C}}_U^L$ of points U.

The accuracy of coordinates of points PL and U surveyed in the area of LN, adjusted in the S(LOC) and transformed into the S(JTSK) will be characterized [with the application of the relevant rules (Anděl 1972; Riečan et al. 1983; Koch 1988)] by covariance matrices:

$$\begin{aligned} \sum\nolimits_{PL}^{Jt} &= s_0^2 \mathbf{A}_{PL} (\mathbf{A}_{PL}^{\mathrm{T}} \mathbf{A}_{PL})^{-1} \mathbf{A}_{PL}^{\mathrm{T}} = s_0^2, \\ \sum\nolimits_{U}^{Jt} &= s_0^2 \mathbf{A}_U (\mathbf{A}_{PL}^{\mathrm{T}} \mathbf{A}_{PL})^{-1} \mathbf{A}_U^{\mathrm{T}}. \end{aligned} \tag{6.53}$$

Coordinate discrepancies (6.49), the size of which is an evident indicator of compatibility of the relevant points PL, are of key importance in the indicated transformation process in terms of verification of compatibility of points PL. If the relevant $\mathbf{V}_X, \mathbf{V}_Y$ for a certain point PL will reach high values, apparently above the expectable stochastic degree of coordinate discrepancies, the relevant point PL determined by coordinates $\mathbf{C}_{PL}^J$ at the epoch t and by coordinates $\mathbf{C}_{PL}^{Jt}$ at the epoch t' will signal its statistically significant change (due to the movement of its physical survey mark during the period $t' - t$, incorrect determination of its coordinates at the epoch t or t', or other reasons). It is natural that the size of $\mathbf{V}_X, \mathbf{V}_Y$ is not assessed subjectively; it is necessary to objectify decisions on accepting or rejecting $\mathbf{V}_X, \mathbf{V}_Y$ by using the appropriate statistical testing procedure.

- **Testing of Coordinate Indicators—Discrepancies**

The used statistical test, of localization character, belongs to the group of so-called tests for outliers (biased values) (Lehmann 2013; Štroner 2014) from a certain set of values, in the present case from a set of coordinates of identical points used for transformation. The test takes into account all influences of measurements, calculations, and transformations in the assessment of statistical significance of the relevant coordinates, or coordinate discrepancies.

The used test may also be considered for its geometric nature as a congruence test of the network structure of points PL in the area of LN from the epoch t and t', while this congruency will always be violated only on the point whose coordinates are significantly different from t and t', i.e., their coordinate indicators are unacceptably high.

In the following parts, when introducing the test, we will highlight only the procedure of its application in accordance with general patterns for the use of statistical test and interpretation of results. In the case of the logical statistical and formal design of the relevant test statistic and its properties, we refer to the literature (Lenzmann 1984; Heck 1985).

The null hypothesis H_0 and alternative hypothesis H_a are formulated for the situation being examined, which, in this case, by their notation:

$$H_0 : \left(d\mathbf{p} = \sqrt{d\mathbf{X}^2 + d\mathbf{Y}^2}\right) = 0, \quad H_a : \left(d\mathbf{p} = \sqrt{d\mathbf{X}^2 + d\mathbf{Y}^2}\right) \neq 0, \quad (6.54a)$$

or in the form:

$$H_0 : \left(\mathbf{V}_\mathbf{C} = \sqrt{V_X^2 + V_Y^2}\right) = 0, \quad H_a : \left(\mathbf{V}_\mathbf{C} = \sqrt{V_X^2 + V_Y^2}\right) \neq 0, \quad (6.54b)$$

express the state of considered point where the vector of incompatibility $d\mathbf{p}$ or its components $d\mathbf{X}, d\mathbf{Y}$ with insignificant (stochastic) values postulate the compatibility of the point (H_0), or with significant values $d\mathbf{p}, d\mathbf{X}, d\mathbf{Y}$ its incompatibility (H_a). A random variable in the following form (Lenzmann 1984) is a test statistic (localization) of every point:

$$T_i = \frac{\frac{R_i}{d}}{\frac{\mathbf{R} - \mathbf{R}_i}{u - \mathrm{tp} - d}} = \frac{u - \mathrm{tp} - d}{d} \frac{\mathbf{R}_i}{\mathbf{R} - \mathbf{R}_i} \sim F(d, u - \mathrm{tp} - d), \quad i = 1, 2, \ldots, p \quad (6.55)$$

that has F-distribution with $f_1 = d, f_2 = u - \mathrm{tp} - d$ degrees of freedom. The meaning of variables and labeling:

p the number of identical (=datum) points PL,

$u = 2p$ the number of coordinates of p points,

$\mathrm{tp} = 4$ the number of transformation parameters,

$d = 2$ coordinate dimension of points,

$$\mathbf{R} = \mathbf{V}_\mathbf{C}^\mathrm{T} \mathbf{Q}_{\mathbf{V}_\mathbf{C}}^{-1} \mathbf{V}_\mathbf{C} \quad (6.56)$$

$$\mathbf{R}_i = \frac{V_{X_i}^2 + V_{Y_i}^2}{1 - \frac{1}{p} - \frac{(X_{r_i}^L)^2 + (Y_{r_i}^L)^2}{\sum_1^p \left((X_{r_i}^L)^2 + (Y_{r_i}^L)^2\right)}} \tag{6.57}$$

where $X_{r_i}^L, Y_{r_i}^L$ are local coordinates reduced to their center of gravity.

The test statistics can be also formulated in the form (Heck 1985):

$$T_i = \frac{\mathbf{R}_i}{d\,\widehat{s}_{0i}^2} \sim F(d, u - \mathrm{tp} - d), \tag{6.58}$$

where

$$\widehat{s}_{0i}^2 = \frac{\mathbf{R} - \mathbf{R}_i}{u - \mathrm{tp} - d}. \tag{6.59}$$

A quadratic form of discrepancies at the point PLi with a unit cofactor matrix of coordinates $\mathbf{C}$ can be determined according to

$$\mathbf{R}_i = \mathbf{V}_{\mathbf{C}}^{\mathrm{T}} \mathbf{H}_i (\mathbf{H}_i^{\mathrm{T}} \mathbf{Q}_{\mathbf{V}_{\mathbf{C}}} \mathbf{H}_i)^{-1} \mathbf{H}_i^{\mathrm{T}} \mathbf{V}_{\mathbf{C}}, \tag{6.60}$$

with a dislocation matrix:

$$\mathbf{H}_i^{\mathrm{T}} = \begin{bmatrix} 0 & 0 & \cdots & 1 & 0 & \cdots & 0 & 0 \\ 0 & 0 & & 0 & 1 & & 0 & 0 \\ & \mathrm{PL}_1 & & & \mathrm{PL}_i & & \mathrm{PL}_p & \end{bmatrix} \tag{6.61}$$

and with a cofactor matrix of discrepancies:

$$\mathbf{Q}_{\mathbf{V}_{\mathbf{C}}} = \mathbf{I} - \mathbf{A}_{PL} (\mathbf{A}_{PL}^{\mathrm{T}} \mathbf{A}_{PL})^{-1} \mathbf{A}_{PL}^{\mathrm{T}}. \tag{6.62}$$

The level of significance α of the test is selected as one of the usual values (6.29).

The critical value of the F-distribution $\{(1 - \alpha_i) - \text{quantile}\}$ for degrees of freedom f_1, f_2 is determined as

$$F_{\alpha_i} = F(\alpha_i; f_1, f_2) \tag{6.63}$$

and values T_i for $i = 1, 2, \ldots, p$ points PL according to (6.55) or (6.58) are compared with F_{α_i} (first stage of verification):

if for the point PLi $T_i < F_{\alpha_i}$,

H_0 is not rejected, and we can be practically confident that the point PLi of superior geodetic control has neither physically nor coordinate significantly changed its position during the period $t' - t$ and it can be considered as compatible in terms of its both components (physical and coordinate position) also at the epoch t' and eligible for the connection of newly established points U at the epoch t';

if $T_i \geq F_{\alpha_i}$,

H_0 is rejected and H_a is apparently true; therefore, a standpoint is adopted (with the risk of incorrect decision α_i) that in the period between epochs t and t' in the physical position of the point PLi of the network LN, or during its survey at the epoch t, there was a significant defect, the formation of incompatibility of point PLi, which cannot be used as a connecting point for determination of new densifying points U now at the epoch t'.

The point PLi (incompatible) is excluded from the set of points PL and the remaining number of $p - 1$ points PL need to be verified once again in terms of their compatibility (second stage of verification).

Elements related to the point PLi are removed from matrices $\mathbf{C}_{PL}^{J}$ (5.45), $\mathbf{A}_{PL}$ (6.44), thus resulting in new matrices $\mathbf{C}_{PL}^{J}$ (2), $\mathbf{A}_{PL}$ (2). By using them, new transformation parameters $\hat{\mathbf{T}}\mathbf{P}$ (2) are determined, transformation $\mathbf{C}_{PL}^{Jt}(2) = \mathbf{A}_{PL}(2)\hat{\mathbf{T}}\mathbf{P}(2)$ is realized, also the relevant coordinate discrepancies $\mathbf{V}_C$ (2) for $p - 1$ points are determined, and then, the appropriate verification of their compatibility using (6.55) or (6.58) and (6.63) is realized:

- if no additional incompatible point is demonstrated, the relevant $p - 1$ compatible points can be used for intended geodetic activities,
- if another incompatible point is identified, it is excluded from $p - 1$ points and the process continues with the remaining $p - 2$ points by analogous phases until the compatibility for all successively remaining points is confirmed.

Example No. 3 In the area of LN of the national planimetric network (ŠTS with the system S-JTSK) with points PL, a densifying geodetic control (points U in the number of $u = 5$) will be established and connected to the selected surrounding points PL of a superior network in the number of $p = 8$ (Fig. 6.5).

Coordinates $\mathbf{C}^{J} = [X\,Y]^{J}$ of points PLi, $i = 1, 2, \ldots, 8$ from the epoch t are as follows (Table 6.7):

It is necessary to ascertain whether points PL are compatible, i.e., if they can be used as datum points for the connection of points U to the ŠTS in the relevant area without the risk that they will incorrectly affect the determination of points U.

New points established at present, at the epoch t', together with points PL create a local network structure LN. A local coordinate system S(LOC) defined by two points, PL1 and PL4 (Fig. 6.5), was established for its survey and coordinate determination only on the basis of measurements (excluding the influence of datum points PL). The network was surveyed and adjusted according to the Gauss–Markov regular model on the required level of accuracy with the connection to points PL1, PL4 with known coordinates of its points that result from the definition of S(LOC) and the accurate determination of the length PL4-PL1.

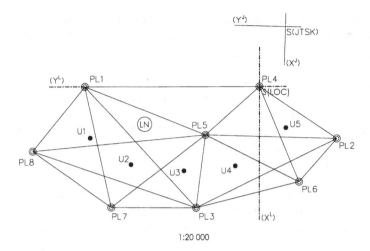

1:20 000

Fig. 6.5 Local network LN of points U and PL in the S(LOC) and S(JTSK). Points U within the LN determined at the epoch t'

For coordinates $\hat{\mathbf{C}}_{PL}^L = [\hat{X}\hat{Y}]_{PL}^L$ of points PL1, ...,PL8 in the system S(LOC), the following values were obtained in Table 6.8: and for coordinates $\hat{\mathbf{C}}_{U}^L = [\hat{X}\hat{Y}]_{U}^L$ of newly determined points $U1$, ...,$U5$ values are shown in Table 6.9:

Coordinates $\hat{\mathbf{C}}_{PL}^L$ are transformed by Helmert transformation into the system S (JTSK) to coordinates $\mathbf{C}_{PL}^{Jt}$ that are valid for points PL1, ..., PL8 at the epoch t' from current measurements.

In the transformation process, using the matrix of coefficients (6.44):

$$
\mathbf{A}_{PL} = \begin{bmatrix}
1 & 0 & 2,000 & -3,210.3896 \\
0 & 1 & 3,210.3896 & 2,000 \\
1 & 0 & 2,358.992 & -1,467.215 \\
0 & 1 & 1,467.215 & 2,358.992 \\
1 & 0 & 2,832.206 & -2,436.892 \\
0 & 1 & 2,436.896 & 2,832.206 \\
1 & 0 & 2,000 & -2,000 \\
0 & 1 & 2,000 & 2,000 \\
1 & 0 & 2,331.44 & -2,371.803 \\
0 & 1 & 2,371.803 & 2,331.44 \\
1 & 0 & 2,654.731 & -1,725.447 \\
0 & 1 & 1,725.447 & 2,654.731 \\
1 & 0 & 2,829.238 & -3,028.401 \\
0 & 1 & 3,028.401 & 2,829.238 \\
1 & 0 & 2,443.675 & -3,573.316 \\
0 & 1 & 3,573.316 & 2,443.675
\end{bmatrix},
$$

Table 6.7 Transformation
parameters

	$\hat{T}P$
dX	1,237,272.357
dY	261,142.0482
m	1.000007345
ω	5.250086865

and transformation parameters determined according to (6.46) (Table 6.10): coordinates $\mathbf{C}_{PL}^{Jt}$ of points PL in the system S(JTSK) were obtained according to (6.47) (Table 6.11): coordinate discrepancies according to (6.49) (Table 6.12): from which vectors of incompatibility $d\mathbf{p}$ results according to (6.54a, 6.54b) (Table 6.13):

Furthermore, the a posteriori unit variance was obtained from the transformation:

$$s_0^2 = 0.0002406, s_0 = 0.0155 \text{ (m)},$$

and according to (6.48) or (6.51) the cofactor, or covariance matrix of transformation parameters (not specified).

Simultaneously, also the coordinates $\mathbf{C}_U^{Jt}$ of determined points U according to (6.52) in the system S(JTSK) were determined by the transformation of coordinates $\hat{\mathbf{C}}_U^L$ (Table 6.8):

According to the results of coordinate discrepancies $\mathbf{V_C}$ (Tables 6.9 and 6.10), we can see that coordinate indicators at the point PL3 indicate significantly outlying (biased) values from the total set of values $\mathbf{V}_X, \mathbf{V}_Y$ and the relevant indicators $\mathbf{V}_X, \mathbf{V}_Y$ at points PL6, PL8 also indicate suspicious outlying values.

Table 6.8 Transformed
coordinates of points U in the
system S(JTSK)

Point	X_U^{Jt} (m)	Y_U^{Jt} (m)
$U1$	1,239,355.190	264,496.6672
$U2$	1,239,559.137	264,230.9993
$U3$	1,239,632.738	263,867.3415
$U4$	1,239,628.997	263,510.2482
$U5$	1,239,397.538	263,140.6541

Table 6.9 Coordinate
discrepancies of points PL

Point	$\mathbf{V}_X.\,dX$ (m)	$\mathbf{V}_Y.\,dY$ (m)
PL1	0.000636	−0.000761
PL2	−0.007222	−0.003732
PL3	−0.022468	0.042006
PL4	0.001373	−0.009423
PL5	0.003683	−0.006623
PL6	0.005907	−0.011366
PL7	0.007751	−0.001842
PL8	0.010342	−0.008262

Table 6.10 Resulting vectors of incompatibility

Point	$V_{p.}.dp$ (m)
PL1	0.000991008
PL2	0.008129275
PL3	0.047637328
PL4	0.009522503
PL5	0.007578167
PL6	0.012809317
PL7	0.007966405
PL8	0.013236979

Localization testing with the Lenzmann–Heck's test (6.55) is realized for an objective assessment of not only those three but of all 8 points PL, so incompatible points can be identified with the required credibility and removed from the set of points PL.

For this purpose, the corresponding quadratic forms of discrepancies are calculated according to (6.56) and (6.57) (Table 6.11):

$$\mathbf{R} = 0.002887258,$$

For the present case, we have $p = 8, u = 16, \text{tp} = 4, r = 2, f_1 = 2, f_2 = 10$, the level of significance of the test $\alpha = 0.01$ is selected, and according to (6.63), we get

$$F_{\alpha_i} = 7.559,$$

and calculated values of T_i for 8 points PL are (Table 6.12):

The following conclusions result from the comparison of T_i and F_{α_i}:

- $T_i > F_{\alpha_i}$ for the point PL3; therefore, H_0 is rejected, with the risk of only $\alpha = 0.01$, i.e., coordinate discrepancies—indicators $dX3, dY3$ at point PL3 can be considered as statistically significant values, clearly outlying from the set of discrepancies of all points PL, with high probability 0.99;

  thus, the point PL3 can be considered as incompatible and not applicable for the determination of points U and all other geodetic activities,

- $T_i < F_{\alpha_i}$, for other points PLi, v H_0 is not rejected, values of coordinate indicators at these points can be considered as stochastic variables. All these points are

Table 6.11 Quadratic form of discrepancies

Point	R_i
PL1	0.000001236
PL2	0.000076747
PL3	0.002839255
PL4	0.000105651
PL5	0.000069361
PL6	0.000195744
PL7	0.000082707
PL8	0.000233024

Table 6.12 Values of test statistics T_i for points PL

Point	T_i
PL1	0.0021
PL2	0.1365
PL3	29.5738
PL4	0.1899
PL5	0.1231
PL6	0.3636
PL7	0.1475
PL8	0.4390

therefore compatible, their verification at the epoch t' confirmed a good consistency between their coordinates $\mathbf{C}_{PL}^{J}$ from the epoch t and $\mathbf{C}_{PL}^{Jt}$ from the epoch t'; these points can be also used for the determination of points U and other geodetic activities in the area of LN.

Based on the above test results, the point PL3 is excluded from the set of 8 PL points and with coordinates of 7 points in the S(LOC) (from Table 6.13) that remained (Tables 6.14 and 6.15):with their coordinates from the S(JTSK) (from Table 6.16):and with coordinates of determined points $U1, \ldots, U5$ in the S(LOC) (from Table 6.17) as well as with the modified structure of matrix $\mathbf{A}_{PL}$ (ignoring lines of the point PL3 from Table 6.18):

$$
\mathbf{A}_{PL} =
\begin{bmatrix}
1 & 0 & 2,000.000 & -3,210.390 \\
0 & 1 & 3,210.390 & 2,000.000 \\
1 & 0 & 2,358.992 & -1,467.215 \\
0 & 1 & 1,467.215 & 2,358.992 \\
1 & 0 & 2,000.000 & -2,000.000 \\
0 & 1 & 2,000.000 & 2,000.000 \\
1 & 0 & 2,331.440 & -2,371.803 \\
0 & 1 & 2,371.803 & 2,331.440 \\
1 & 0 & 2,654.731 & -1,725.447 \\
0 & 1 & 1,225.447 & 2,654.731 \\
1 & 0 & 2,829.238 & -3,028.401 \\
0 & 1 & 3,028.401 & 2,829.238 \\
1 & 0 & 2,443.675 & -3,573.316 \\
0 & 1 & 3,573.316 & 2,443.675
\end{bmatrix},
$$

the new determination of $\hat{\mathbf{TP}}$ and new transformation is realized with the following results:

- transformation parameters (Table 6.19):
- transformed coordinates $\hat{\mathbf{C}}_{PL}^{L}$ to $\mathbf{C}^{Jt}$ (Table 6.20):
- coordinate discrepancies—indicators (Table 6.21):

Table 6.13 Coordinates of points PL in the S(LOC) system

Point	$\hat{X}^L$ (m)	$\hat{Y}^L$ (m)
PL1	2000.000	3210.390
PL2	2358.992	1467.215
PL3	2832.206	2436.892
PL4	2000.000	2000.000
PL5	2331.440	2371.803
PL6	2654.731	1725.447
PL7	2829.238	3028.401
PL8	2443.675	3573.316

Table 6.14 Coordinates of remaining PL points in the S (LOC)

Point	$\hat{X}_{PL}^L$ (m)	$\hat{Y}_{PL}^L$ (m)
PL1	2000.000	3210.390
PL2	2358.992	1467.215
PL4	2000.000	2000.000
PL5	2331.440	2371.803
PL6	2654.731	1725.447
PL7	2829.238	3028.401
PL8	2443.675	3573.316

Table 6.15 Coordinates of remaining PL points in the S (JTSK)

Point	X_{PL}^J (m)	Y_{PL}^J (m)
PL1	1,239,001.117	264,506.302
PL2	1,239,502.494	262,798.614
PL4	1,239,100.823	263,300.026
PL5	1,239,400.509	263,697.868
PL6	1,239,775.945	263,080.340
PL7	1,239,842.527	264,393.240
PL8	1,239,413.382	264,904.553

Table 6.16 Coordinates of points PL*i*

Point	X^J (m)	Y^J (m)
PL1	1,239,001.117	264,506.302
PL2	1,239,502.494	262,798.614
PL3	1,239,894.241	263,803.938
PL4	1,239,100.823	263,300.026
PL5	1,239,400.509	263,697.868
PL6	1,239,775.945	263,080.340
PL7	1,239,842.527	264,393.240
PL8	1,239,413.382	264,904.553

Table 6.17 Coordinates of points U in the S(LOC) system

Point	$\hat{X}^L$ (m)	$\hat{Y}^L$ (m)
$U1$	2352.073	3171.622
$U2$	2533.441	2890.059
$U3$	2576.835	2521.577
$U4$	2543.692	2166.008
$U5$	2282.576	1816.739

Table 6.18 Transformed coordinates of points PL in the system S(JTSK)

Point	X_{PL}^{Jt} (m)	Y_{PL}^{Jt} (m)
PL1	1,239,001.118	264,506.301
PL2	1,239,502.487	262,798.610
PL3	1,239,894.219	263,803.980
PL4	1,239,100.824	263,300.016
PL5	1,239,400.512	263,697.861
PL6	1,239,775.951	263,080.328
PL7	1,239,842.535	264,393.239
PL8	1,239,413.392	264,904.545

Table 6.19 Resulting transformation parameters

	$\hat{\text{T}}\text{P}$
dY (m)	1,237,272.371
dY (m)	261,142.051
m	1.000004943
$\omega[^{cc}]$	5.250361746

Table 6.20 Transformed coordinates of PL points

Point	X^{Jt} (m)	Y^{Jt} (m)
PL1	1,239,001.112	264,506.304
PL2	1,239,502.488	262,798.619
PL4	1,239,100.824	263,300.022
PL5	1,239,400.511	263,697.867
PL6	1,239,775.950	263,080.337
PL7	1,239,842.528	264,393.245
PL8	1,239,413.384	264,904.548
PL1	1,239,001.112	264,506.304

Table 6.21 Coordinate discrepancies

Point	$\mathbf{V}_X$ (m)	$\mathbf{V}_Y$ (m)
PL1	−0.004535	0.001704
PL2	−0.006223	0.004998
PL4	0.001171	−0.003632
PL5	0.001044	−0.000493
PL6	0.005033	−0.002131
PL7	0.001048	0.004528
PL8	0.002462	−0.004975
PL1	−0.004535	0.001704

Table 6.22 Vectors of incompatibility

Point	V_p (m)
PL1	0.00484
PL2	0.00798
PL4	0.00382
PL5	0.00115
PL6	0.00547
PL7	0.00465
PL8	0.00555
PL1	0.00484

Table 6.23 Transformed coordinates of U points

Point	X_U^J (m)	Y_U^J (m)
U1	1,239,355.190	264,496.667
U2	1,239,559.137	264,230.999
U3	1,239,632.738	263,867.342
U4	1,239,628.997	263,510.248
U5	1,239,397.538	263,140.654

- vectors of incompatibility (Table 6.22):
- a posteriori unit variance:

$$s_0^2 = 0.000018535, s_0 = 0.00431 \ [\text{m}],$$

- and coordinates C_U^J based on the transformation of $\hat{C}_U^L$ into the S(JTSK) according to (6.52) (Table 6.23):

As can be seen from the numerical values of coordinate discrepancies—indicators, discrepancies V_X, V_Y (Table 6.21) at all points represent a stochastic size even without their testing, points are therefore compatible.

However, formally, it is necessary to continue with a testing of the point that has the highest values of V_X, V_Y (point PL2 from Table 6.21) also in this set of 7 points. If the result of this test will be $T_{PL2} < F_{\alpha_i}$, then it is possible to adopt the standpoint of this set of 7 points that these points are compatible.

Thus, in the present case, only 7 points without the incompatible point PL3 can be used to determine new points U or for geodetic activities in the area of LN from the set of verified points PL1, PL2, PL3, PL4, PL5, PL6, PL7, PL8.

Koch's Method

All those information, situational and transformational data (also the Helmert transformation) are used in the application of this testing method for verifying the compatibility of geodetic control of PL points as in the use of Lenzmann–Heck's method for the Example No. 3 (Sect. Lenzmann–Heck's Method, Fig. 6.5). Thus, also the use of this examination of the character of planimetric points will be presented by a verification of compatibility of points PL1, ..., PL8 with all data and characteristics as well as for the same reasons and objectives as Sect. Lenzmann–Heck's Method.

- **Testing of Coordinate Indicators—Discrepancies**

Also this featured procedure according to Koch belongs to the group of statistical tests for the identification of outlying values from a certain file (Koch 1983; Kok 1984) and in the present case, it is used for coordinates of datum points PL (used as identical points for transformation), or coordinate discrepancies $\mathbf{V}_C$ created from them.

Testing by this test is usually done only on the level of localization testing (Koch 1983, 1985), in which the individual points PL are separately verified for compatibility.

Variables that will be available from the epoch t (the formation of superior geodetic control) and from the epoch t' (the current survey of local geodetic control with points PL and newly determined points U) appear at the test statistics for localization tests:

p the number of datum (identical) points,

$u = 2p$ the number of coordinates of p points,

tp   the number of transformation parameters,

$d = 2$ coordinate dimension,

$$\mathbf{C}^J_{PL} = [X_1 \; Y_1 \; \ldots \; X_8 \; Y_8]^J_{PL} \tag{6.64}$$

- S-JTSK coordinates of 8 selected datum points PL from a superior network (from the epoch t),

$$\mathbf{V_C} = \begin{bmatrix} V_{X1} & V_{Y1} \\ \vdots & \vdots \\ V_{X8} & V_{Y8} \end{bmatrix} = \mathbf{C}^J_{PL} - \mathbf{A}_{PL}\hat{\mathbf{T}}\mathbf{P} = \mathbf{C}^J_{PL} - \mathbf{C}^{Jt}_{PL} \tag{6.65}$$

- coordinate discrepancies, indicators identified for points PL within transformation ("corrections" of coordinates of identical points in the determination of transformation parameters),

$$\mathbf{V}_{Ci} = \begin{bmatrix} v_{Xi} \\ v_{Yi} \end{bmatrix} \tag{6.66}$$

- a vector of coordinate discrepancies relating to the ith point of PL points,

$\mathbf{A}_{PL}$ a matrix of coefficients with coordinates $\hat{\mathbf{C}}_{PL}^{L}$ or their reduced values with the structure according to (6.44),

$$\mathbf{Q}_{\mathbf{V_C}} = \mathbf{I} - \mathbf{A}_{PL}(\mathbf{A}_{PL}^{\mathrm{T}}\mathbf{A}_{PL})^{-1}\mathbf{A}_{PL}^{\mathrm{T}} \tag{6.67}$$

- $n \times n$ matrix of cofactors of coordinate discrepancies,

$$s_0^2 = \frac{\mathbf{V}_{\mathbf{C}}^{\mathrm{T}}\mathbf{V_C}}{n - \mathrm{tp}} \tag{6.68}$$

- a posteriori unit variance from the determination of transformation parameters,
  $\mathbf{Q_L}$ a cofactor matrix of observations at t'

Test statistic for the localization test is used in the form (Koch 1985):

$$T_i = \sqrt{\frac{\mathbf{R}}{\mathrm{d}.s_0^2}} \sim \tau_{\alpha_0} \tag{6.69}$$

where the quadratic form of point coordinate discrepancies is defined as

$$\mathbf{R}_i = \mathbf{V}_{\mathbf{C}}^{\mathrm{T}}\mathbf{Q_L}^{-1}\mathbf{H}_i(\mathbf{H}_i^{\mathrm{T}}\mathbf{Q_L}^{-1}\mathbf{Q}_{\mathbf{V_C}}\mathbf{Q_L}^{-1}\mathbf{H}_i)^{-1}\mathbf{H}_i^{\mathrm{T}}\mathbf{Q_L}^{-1}\mathbf{V_C} \tag{6.70}$$

and for $\mathbf{Q_L} = \mathbf{I}$:

$$\mathbf{R}_i = \mathbf{V}_{\mathbf{C}}^{\mathrm{T}}\mathbf{H}_i(\mathbf{H}_i^{\mathrm{T}}\mathbf{Q}_{\mathbf{V_C}}\mathbf{H}_i)^{-1}\mathbf{H}_i^{\mathrm{T}}\mathbf{V_C}, \tag{6.71}$$

with the localization matrix $\mathbf{H}_i$ of point PLi (6.61). For every ith point PLi, the relevant $\mathbf{R}_i$ can be separately determined also approximately according to

$$\mathbf{R}_i = \mathbf{V}_{Ci}^{\mathrm{T}}\mathbf{Q}_{\mathbf{V}_{Ci}}^{-1}\mathbf{V}_{Ci}. \tag{6.72}$$

The test statistic T_i has τ-probability distribution, whose quantiles are determined using the F-distribution according to (Koch 1985):

$$\tau_{\alpha_0} = \sqrt{\frac{(u - \mathrm{tp})F_{\alpha_0}}{u - \mathrm{tp} - d + dF_{\alpha_0}}}, \tag{6.73}$$

where

$$F_{\alpha_0} = F(1 - \alpha_0; f_1, f_2),$$
$$\alpha_0 = \frac{\alpha}{p}, f_1 = d, f_2 = u - \text{tp} - d. \tag{6.74}$$

Two different, in terms of compatibility, possible situations of coordinate discrepancies $\mathbf{V}_X, \mathbf{V}_Y$ at individual points PL are again formulated in the form of null hypothesis and the corresponding alternative hypothesis for the localization test:

$$H_0 : [V_{Xi}, V_{Yi}] = 0, \ H_a : [V_{Xi}, V_{Yi}] \neq 0, \tag{6.75}$$

of which the H_0 postulates an opinion that the vector $\mathbf{V}_{Ci}$ is a random vector with insignificant, stochastic values V_{Xi}, V_{Yi} for the relevant point PLi, i.e., that this point can be considered as compatible. The hypothesis H_a represents the opposite view that the realized vector $\mathbf{V}_{Ci}$ contains non-stochastic, statistically significant values V_{Xi}, V_{Yi} at the relevant point PLi, which can, therefore, be considered as incompatible.

If the following results for a certain point PLi in the testing:

$$T_i < \tau_{\alpha_0},$$

H_0 is not rejected and PLi can be considered as a compatible point. If

$$T_i \geq \tau_{\alpha_0},$$

H_0 is rejected, the relevant point will be legitimately declared as incompatible, and it is, therefore, necessary to exclude it from the geodetic control of PL points, in order to be not used for geodetic activities.

Example No. 4 The geodetic control with the situation, data and problems specified in the Example No. 3, Fig. 6.5 and all the solutions and preparatory calculations for the corresponding test are used to demonstrate the introduced test method:

$$p = 8,$$
$$\text{tp} = 4,$$
$$u = 2p = 16,$$
$$d = 2,$$
$$f_1 = d, f_2 = n - \text{tp} - d,$$
$$s_0^2 = 0.0002406, \ s_0 = 0.0155 \, (\text{m})$$

and also identical matrices $\mathbf{C}^J, \mathbf{A}_{\text{PL}}, \mathbf{V}_X, \mathbf{V}_Y$ as specified in the Example No. 3, Sect. Lenzmann–Heck's Method.

The cofactor matrix of coordinate discrepancies $\mathbf{Q_{v_c}}$ (not specified) was determined according to (6.67) and subsequently, values of quadratic forms of point coordinate discrepancies for points PL1, ..., PL8 were determined in terms of (6.71) with the following values (Table 6.24).

Null and alternative hypotheses in the testing of this method are formulated as in Sect. Lenzmann–Heck's Method and $\alpha_0 = 0.01$ is used for the level of significance.

The identification of incompatible points in the geodetic control PL is then realized on the basis of localization testing, in which every point PLi is tested separately using the appropriate relations (6.69), (6.73), and (6.74) (Table 6.25).

Realizations of test statistic T (6.69) for individual points give following values: and according to (6.73) and (6.74), values $F_{\alpha_0}, \tau_{\alpha_0}$ for testing are

$$F_{\alpha_0} = F(0.01, \ 2, \ 16 - 4 - 2) = 7.559,$$

$$\tau_{\alpha_0} = \sqrt{\frac{(16-4)7.559}{16-4-2+7.559}} = 1.9003.$$

Table 6.24 Quadratic form of discrepancies

Point	$\mathbf{R}_i$
PL1	1.365562E−6
PL2	0.000101122
PL3	0.002701886
PL4	0.000115420
PL5	0.000065975
PL6	0.000221068
PL7	0.000081886
PL8	0.000284922

Table 6.25 Values of test statistics

Point	T_i
PL1	0.05327
PL2	0.45841
PL3	2.36956
PL4	0.48975
PL5	0.37028
PL6	0.67779
PL7	0.41252
PL8	0.76948

As results from the confrontation of T_i with τ_{α_0}, for the point PL3,

$$PL3 : T_i > \tau_{\alpha_0},$$

and for other points,

$$PL1, PL2, PL4, PL5, PL6, PL7, PL8 : T_i < \tau_{\alpha_0}.$$

Thus, the following results from localization testing: we are eligible to reject H_{0i} for the point PL3 (with the risk of incorrect rejection α_0), because coordinate discrepancies at the point are clearly significant, i.e., the point PL3 is an incompatible point whose coordinate positions at t and t' are not stochastically identical.

It results from the testing of other points that H_{0i} are not rejected for these points, coordinate discrepancies at all points have a stochastic character, i.e., all points can be considered as compatible and, therefore, they are suitable as connecting (datum) points for the determination of new points U.

6.3   The Compatibility Verification with dL Indicators

6.3.1   In General

As stated in Chap. 3, it is theoretically possible to use values of dC and dL to examine the compatibility of 2D points. The characteristics of these indicators, possibilities of their applications, advantages, and disadvantages were specified in Sects. 3.4.4 and 3.4.5, as well as the conclusion that, at present, the preference is given to indicators—coordinate differences $d\mathbf{C} = [dX, dY]$, which directly inform about examined characteristics of points and by which, their compatibility characteristics can be clearly characterized with a high degree of credibility.

Indicators $d\mathbf{L} = [dd, d\omega, \ldots]$ themselves, identified in a geodetic control and without the assistance of dC indicators, cannot provide (or only by complicated analyzes) clear conclusions regarding the compatibility of a particular point in many cases. The causes were also pointed out in Sects. 3.4.4 and 3.4.5 and their lack of applicability was as well graphically demonstrated (e.g., in Fig. 6.5).

Therefore, indicators dL in current practice are independently used most commonly in the geometric assessment of compatibility characteristics of one or very few points (determined by oriented distances, intersections, resections, etc.). So far as they are used for network structures in a geodetic control, the use of dL is equivalent to dC indicators, as follows from the proof of equivalence of their quadratic forms.

Following model relations apply in the network structure:

$$\mathbf{L} = f(\mathbf{C}, \ldots),$$

in which, if coordinates $\mathbf{C}$ are changed by values $d\mathbf{C}$, also $\mathbf{L}$ will change by values $d\mathbf{L}$, in terms:

$$\mathbf{L} + d\mathbf{L} = f(\mathbf{C} + d\mathbf{C}, \ldots).$$

For $d\mathbf{L}$ from the Taylor series expansion, by neglecting terms of the first and higher orders, we get

$$d\mathbf{L} = f(\mathbf{C}) + \frac{\partial \mathbf{L}}{\partial \mathbf{C}} d\mathbf{C} - \mathbf{L} = \frac{\partial \mathbf{L}}{\partial \mathbf{C}} d\mathbf{C} = \mathbf{F} d\mathbf{C} \qquad (*)$$

where $\mathbf{F}$ is the Jacobian of the vector $d\mathbf{L}$ with regard to the vector $\mathbf{C}$. The cofactor matrix of the vector $d\mathbf{L}$ is defined as

$$\mathbf{Q}_{d\mathbf{L}} = \mathbf{F} \mathbf{Q}_{d\mathbf{C}} \mathbf{F}^{\mathrm{T}}, \qquad (**)$$

where $\mathbf{Q}_{d\mathbf{C}}$ is the cofactor matrix of the vector $d\mathbf{C}$.

Quadratic form of the vector $d\mathbf{L}$ is defined as

$$\mathbf{R}_{d\mathbf{L}} = d\mathbf{L}^{\mathrm{T}} \mathbf{Q}_{d\mathbf{L}}^{-1} d\mathbf{L}, \qquad (6.76)$$

and after the substitution for $d\mathbf{L}$ according to $(*)$ and $(**)$:

$$\mathbf{R}_{d\mathbf{L}} = (\mathbf{F} d\mathbf{C})^{\mathrm{T}} (\mathbf{F} \mathbf{Q}_{d\mathbf{C}} \mathbf{F}^{\mathrm{T}})^{-1} (\mathbf{F} d\mathbf{C}) = d\mathbf{C}^{\mathrm{T}} \mathbf{Q}_{d\mathbf{C}}^{-1} d\mathbf{C} = \mathbf{R}_{d\mathbf{C}}, \qquad (6.77)$$

and it results from this equality that quadratic forms of changes of observed variables $d\mathbf{L}$ and coordinate changes $d\mathbf{C}$ are numerically identical.

In the application of $d\mathbf{L}$, it should be pointed out that preference is given to length differences $dd_{ij} = d_{ij} - d'_{ij}$ of various types of measured geometrical variables, apparently for reasons of simplicity of solutions and interpretations. It should also be emphasized that all methods and procedures for verification of compatibility of points relate only to point situations with full pre-information.

Following the mentioned, the verification of changes in lengths between points and applicable conclusions for compatibility of points resulting from them (Welsch 1983; Polak 1984; Stichler 1981) will be mentioned in the next part on a symbolic level. Applications of $d\mathbf{L}$ or $d\mathbf{C}$ and $d\mathbf{L}$ in various geodetic point structures and from different perspectives can also be found in the works (Stichler 1982, 1985; Werner 1983 and others).

6.3.2 The Verification of Compatibility of Points of Changes in Lengths

6.3.2.1 Verification on the Basis of One Length

Let the set of points with p points P be surveyed in trilateration structure and adjusted at the epoch t and analogously also at the epoch t'.

Let the length with standard deviation is determined between points P_i and P_j (Fig. 6.6) with coordinates $C_i = [\hat{X}_i \, \hat{Y}_i]$, $C_j = [\hat{X}_j \, \hat{Y}_j]$ at epochs t and t', with values:

$$\hat{d}_{ij}, \; s_{\hat{d}_{ij}},$$
$$\hat{d}'_{ij}, \; s_{\hat{d}'_{ij}}, \tag{6.78}$$

that determine the length indicator:

$$dd_{ij} = \hat{d}_{ij} - \hat{d}'_{ij}, \tag{6.79}$$

and its standard deviation:

$$s_{dd_{ij}} = \sqrt{s_{\hat{d}_{ij}}^2 + s_{\hat{d}'_{ij}}^2}. \tag{6.80}$$

Based on the statistical significance of dd_{ij}, it is possible to assess whether the length d_{ij} has been significantly changed or not during the period $t' - t$, i.e., whether the position of one endpoint P_i or P_j or both points has changed.

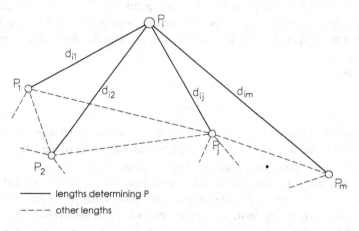

Fig. 6.6 Point P_i in a trilateration network determined by measured lengths

The relevant null and alternative hypothesis:

$$H_0 : dd_{ij} = 0, \; H_a : dd_{ij} \neq 0 \tag{6.81}$$

postulates that dd_{ij} is statistically insignificant stochastic value (H_0) or that dd_{ij} should be considered as statistically significant value (H_a). For example, the following statistics (Welsch 1983) can be used for testing (6.79) on the selected level of significance α:

$$T = \frac{dd_{ij}^2}{s_{dd_{ij}}^2} = \frac{dd_{ij}^2}{\bar{s}_0^2 \cdot f} \sim F(1, n - u) \tag{6.82}$$

that has the F-distribution with $f_1 = 1$ and $f_2 = n - u$ degrees of freedom, where $\bar{s}_0^2$ is the common a posteriori unit variance (6.23) from both adjustments at t and t', n is the number of measured lengths in the trilateration network at t and t', $u = 2p$ is the number of determined parameters—coordinates of points P.

The critical value of the F-distribution $((1 - \alpha)$-quantile) is

$$T_\alpha = F(\alpha; 1, f = n - u). \tag{6.83}$$

If the following results from the comparison of T, T_α:

$$T < T_\alpha,$$

H_0 is not rejected, the tested length d_{ij} remains unchanged within the accuracy of its determination at the time $t' - t$, i.e., its value d'_{ij} at t' differs from d_{ij} only in insignificant stochastic changes resulting from measurements. This conclusion further follows that the endpoints P_i, P_j of length have also changed only insignificantly during the period $t' - t$, i.e., points P_i, P_j can be considered as compatible at the epoch t' (if the possibility that both points have moved significantly, but the length d'_{ij} between them should not significantly differ from d_{ij}, is excluded).

However, if

$$T \geq T_\alpha,$$

H_0 is rejected, meaning that the length d_{ij} has significantly changed at the time $t' - t$ (to the value d'_{ij}), which further implies that one of the points P_i, P_j or both points have also significantly changed their position.

It is necessary to use additional verification procedures, especially with $d\mathbf{C}$ indicators, to determine changes of a specific point (changes of its coordinates) since the used $d\mathbf{L} \equiv dd_{ij}$ indicator cannot provide clear standpoint for the conclusion of this type.

6.3.2.2 Verification Based on a Set of Lengths

Let the point P_i from the set of p points P, which should be verified for compatibility, be determined by the number of m measured lengths $d_{i1}, d_{i2}, \ldots, d_{im}$ determining the point P_i from other points of the network (Fig. 6.2) within the structure of trilateration network at the epoch t'. Point P_i is determined by corresponding lengths $d'_{i1}, d'_{i2}, \ldots, d'_{im}$ at the epoch t'. The network was adjusted at each epoch and from the results of monitored objective: Adjusted values of lengths, their cofactor matrices, and a posteriori unit variances are used:

$$t : \hat{d}, \ \mathbf{Q}_{\hat{d}}, \ s_0^2,$$
$$t' : \hat{d}', \ \mathbf{Q}_{\hat{d}'}, \ s_0'^2, \tag{6.84}$$

and subsequently, the following are determined from them:

– observational indicators $d\mathbf{L} \equiv dd$:

$$dd = \hat{d} - \hat{d}', \tag{6.85}$$

– cofactor matrix of differences dd:

$$\mathbf{Q}_{dd} = \mathbf{Q}_{\hat{d}} + \mathbf{Q}_{\hat{d}'}, \tag{6.86}$$

– a common a posteriori unit variance in terms of (6.23):

$$\bar{s}_0^2 = \frac{\mathbf{v}_d^T \mathbf{Q}_d^{-1} \mathbf{v}_d + \mathbf{v}_{d'}^T \mathbf{Q}_{d'}^{-1} \mathbf{v}_{d'}}{(n-u) + (n-u)}. \tag{6.87}$$

Using length differences—indicators, it is possible to assess the changelessness, or changes in lengths between points P_i and $P_1, \ldots, P_m$ during the period $t' - t$ and, therefore, adopt certain standpoints on the positional situation of these points.

The null hypothesis is formulated for testing length changes:

$$H_0 : \Sigma(dd)^2 = 0, \tag{6.88}$$

assuming that none of m lengths determining point P_i has significantly changed during the period $t' - t$, against the alternative hypothesis:

$$H_a : \Sigma(dd)^2 \neq 0. \tag{6.89}$$

The following test statistic (Welsch 1983) is used to verify H_0:

$$T_i = \sum_1^m \sum \left(\frac{dd}{s_{dd}}\right)^2 = \sum_1^m \sum \left(\frac{dd}{\bar{s}_0 \sqrt{q_{dd}}}\right)^2 \sim F(m, f = 2(n-u)) \tag{6.90}$$

where the cofactor for dd lengths d_{ij} based on (6.86) is defined as

$$q_{dd_{ji}} = q_{\hat{d}_{ji}} + q_{\hat{d}'_{ji}},$$ (6.91)

and:

m is the number of lengths determining the point P_i,
n is the number of all lengths in one network realization,
u is the number of coordinates of points in the network.

Testing is carried out on the level of significance α (6.29), thus $(1 - \alpha)$-quantile of the F-distribution is

$$F_{\alpha_i} = F(\alpha; m, f = 2(n - u)).$$ (6.92)

If the following results from the comparison of T and F_{α_i}:

$$T_i < F_{\alpha_i},$$

H_0 is not rejected; therefore, none of the tested lengths between points $P_1 P_i$, ..., $P_m P_i$ have statistically significantly changed its size during the period $t' - t$ and consequently, points determining these lengths have not significantly changed their positions. Points $P_i, P_1, ..., P_m$ can, therefore, be considered as compatible points.
   In the case of

$$T_i \geq F_{\alpha_i},$$

the result indicates statistically significant changes in some length(s), whereby it is not possible to judge about specific changed lengths and specific points, whose compatibility we are interested in. The test detects only that of the total number of network points, some point(s) can be positionally significantly changed or incorrectly determined, i.e., incompatible, in the set of $m + 1$ points. In that case, various additional verifications and tests are necessary.

6.3.3   The Verification of Compatibility of Points in Network Structure

In the 2D trilateration network (Fig. 6.7), lengths between points are measured at the epoch t and t' and adjusted values $\hat{d}, \hat{d}'$ of measured lengths as well as adjusted values of coordinates $\hat{\mathbf{C}} = [\hat{X}\hat{Y}]$, $\hat{\mathbf{C}}' = [\hat{X}'\hat{Y}']$ of network points in the system $S(XY)$ are determined on the basis of adjustment of the network.
   The compatibility of network points can be verified by various methods (Polak 1984; Welsch 1983; Czaja 1996), of which the procedure using length differences

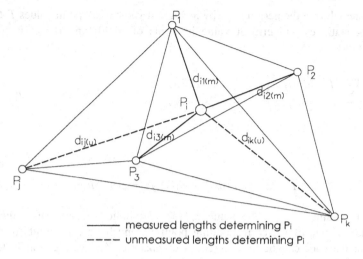

$$\text{------ measured lengths determining P}_i$$
$$\text{- - - - unmeasured lengths determining P}_i$$

Fig. 6.7 Point P_i in the trilateration network determined by measured $d(m)$ and unmeasured d (u) lengths

—indicators *dd* for measured and unmeasured lengths (Welsch 1983) will be introduced.

The verification procedure is realized in cycles using the test statistic (6.90), in which also indicators *dd* of unmeasured lengths $d(u)$ are used in addition to indicators *dd* for measured lengths $d(m)$, all of them directed at the verified point. Length discrepancies *dd* are determined from adjusted coordinates of endpoints, for example, for the length (Fig. 6.7) between points P_i and P_j:

$$dd_{ij}(u) = \hat{d}_{ij} - \hat{d}'_{ij} = \sqrt{(\hat{X}_j - \hat{X}_i)^2 + (\hat{Y}_j - \hat{Y}_i)^2} - \sqrt{(\hat{X}'_j - \hat{X}'_i)^2 + (\hat{Y}'_j - \hat{Y}'_i)^2}$$

$$(6.93)$$

Thus, also differences $dd(u)$ from unmeasured lengths $d_{ij}(u)$, $d_{ik}(u)$ are used for point P_i that is surveyed by lengths $d_{i1}(m)$, $d_{i2}(m)$, $d_{i3}(m)$ and of which differences $dd(m)$ are derived.

The algorithm for the identification of incompatible points is realized as follows.

1st Cycle

Realizations of the statistic (6.90) with corresponding values of variables for every point of the network are determined by the outlined method. Thus, we obtain the following values for individual points:

$$T_1, T_2, T_3, \ldots, T_i, T_j, T_k,$$

of which, let the T_i gives the maximum value. With this test statistics T_i, that is the most suspicious of the fact that the relevant point P_i has changed positionally (or by coordinates) during the period $t' - t$ by its highest value of all point values T, testing by using statistics and critical value in terms of (6.90) and (6.92) is realized according to

$$T_i = \sum_1^{n(dm)} \left(\frac{dd(m)}{\bar{s}_0 \sqrt{q_{dd(m)}}} \right)^2 + \sum_1^{n(dn)} \left(\frac{dd(u)}{\bar{s}_0 \sqrt{q_{dd(n)}}} \right)^2 \sim F(n(dm) + n(du), 2(n - u))$$

(6.94)

and

$$F_{\alpha_i}(\alpha; f_1 = n(dm) + n(du), f_2 = 2(n - u)).$$ (6.95)

For the result $T_i < F_{\alpha_i}$, the value T_i does not indicate statistically significant positional change of the corresponding point P_i, and hence, all points in the trilateration network can be considered as compatible. The verification is thereby completed.

If there is the result of $T_i \geq F_{\alpha_i}$, the corresponding T_i can be considered as statistically significant value and hence the relevant point P_i as incompatible, for which significant positional or coordinate changes occurred over the period $t' - t$. Point P_i is therefore excluded from the further procedure, and remaining points are once again tested in the 2nd cycle, to make the possible another incompatible point be identified.

2nd Cycle

The corresponding values of statistics according to (6.94) are determined for remaining points of network:

$$T_1, T_2, T_3, \ldots, T_j, T_k,$$

of which let the point P_3 have the maximum numerical value with the value of T_3. The corresponding critical value F_{α_3} is determined in terms of (6.95) for this point, and the testing is once again carried out with realizations T_3, F_{α_3}. If

$$T_3 < F_{\alpha_3},$$

the relevant values of T (from the 1st cycle) will be of a stochastic character for the point P_3 and thereby for all other points, and, therefore, all 5 remaining points can be considered as compatible; but if

$$T_3 \geq F_{\alpha_3},$$

it indicates a situation where the value of T_3 on the point P_3 is statistically significant among 5 points, i.e., the relevant point P_3 is incompatible. The point P_3 is excluded, and the remaining 4 points P_1, P_2, P_j, P_k are further verified by analogous procedures within additional test cycles.

The algorithm continues until the kth cycle, in which the result $T_j < F_\alpha$ (i.e., the situation where (in the kth cycle) none of the tested T represents non-stochastic value) is identified by the test (6.94), (6.95) for the highest value T_j of the remaining values T from the previous $(k-1)$-cycle.

Chapter 7
The Verification of Compatibility of Height Points

7.1 In General

The compatibility of height points HL, i.e., a geometric height identity of horizontal tangent plane of height survey mark (Fig. 4.1) with height data h from a certain vertical datum $S(h)$ defined relative to the geoid or quasigeoid by the relevant level surface of the field of gravity (tangential to the survey mark), is just as important for the correct and reliable performance of altimetric measurements and determination of heights of new points as requirements for positional compatibility of geodetic controls. The same causes (physical subsidence of survey mark, the influence of datums, incorrect determination of height, etc.), which may violate the required compatibility and deteriorate the current determination of heights, also apply to height points. Therefore, it is always necessary to verify the state of connecting (datum) height points in terms of their compatibility before important altimetric measurements, so that only reliable, credible, and mutually compatible connecting points could be used for altimetric connection of new points.

Same principles as for the examination of planimetric compatibility also apply to the examination of altimetric compatibility, in particular:

- the compatibility is examined on the basis of two determinations of heights of verified points of different time;
- the condition of height variable in time of the same point is assessed on the basis of the theory of stochastic phenomena with the apparatus of mathematical statistics.

© The Author(s) 2016
G. Weiss et al., *Survey Control Points*,
SpringerBriefs in Geography, DOI 10.1007/978-3-319-28457-6_7

7.2   Strategies of Verification of Altimetric Compatibility

Regarding needs of pre-information necessary to examine the altimetric compatibility, we have two different situations also in this case:

- a situation in which all the necessary data and values are available for verification (full pre-information) from both epochs t and t' and
- a situation in which only heights h from the epoch t and all the necessary data for verification from the epoch t' are known (incomplete pre-information).

7.2.1   The Verification with Full Pre-information

Analogously to Sect. 6.2.1.1 for planimetric points, all the necessary data from measurements and adjustment at epochs t and t' must be available also for the verification of height points. These data represent mainly:

$\hat{\mathbf{h}}, \hat{\mathbf{h}}'$	adjusted heights of verified points from both epochs,
$\mathbf{Q}_{\hat{\mathbf{h}}}, \mathbf{Q}_{\hat{\mathbf{h}}'}$	cofactor matrices of adjusted heights,
$k = k'$	the number of determined height points at t and t',
n, n'	the number of measured differences in elevation between points at t and t',
$s_0^2 = \frac{\Omega}{n-k}$	a posteriori unit variance from the adjustment at t, and
$s_0'^2 = \frac{\Omega'}{n'-k'}$	a posteriori unit variance from the adjustment at t'.

Based on these, it is possible to realize the verification of altimetric compatibility for this situation by suitable methods (Lenzmann 1984; Jakub 2001 and others), of which the so-called method of altimetric congruence will be specified (Niemeier 1980; Jakub 2001) in Sect. 7.3.1.

7.2.2   The Verification with Incomplete Pre-information

To verify the altimetric compatibility of level network in this case, only the following data from epochs t and t' are available:

$\mathbf{h}(\hat{\mathbf{h}}), \hat{\mathbf{h}}'$	adjusted heights of verified points from 2 epochs,
$\mathbf{Q}_{\hat{h}'}$	cofactor matrices of adjusted heights from t',
$\mathbf{R}'$	quadratic form of height discrepancies from t',
k'	the number of height points at t',
n'	the number of measured differences in elevation at t', and

$s_0'^2$ a posteriori unit variance from the adjustment at t'.

Since only vectors **h** and **h'** can be verified in this case, but it is not possible to assess their differences $d\mathbf{h} \equiv \mathbf{V_h}$ in terms of stochastic characteristics of both measurements and their adjustments, it is generally approached at least to the a priori elimination of potential influence of different datums in $d\mathbf{h}$, so that these discrepancies realistically represent influences and effects only of other factors (especially a displacement of height point, incorrect determination of height). Also for these observations, Helmert transformation of values $\hat{\mathbf{h}}'$ to $\mathbf{h}^t$ and the formation of height discrepancies $\mathbf{V_h}$ from the transformation as well as their cofactor matrix are used for it according to:

$$\mathbf{V_h} = \hat{\mathbf{h}}' - \mathbf{h}^t, \quad \mathbf{Q_{V_h}} = \mathbf{Q_{\hat{h}'}} + \mathbf{Q_{h'}} = \mathbf{Q_{\hat{h}'}} + \mathbf{A Q_{\hat{T}P} A^T}, \tag{7.1}$$

where $\mathbf{Q_{\hat{T}P}}$ is the cofactor matrix of determined transformation parameters and **A** is the matrix of heights **h'** (suitably reduced) with the usual structure.

Even in that case of incomplete pre-information, it is possible to verify the altimetric compatibility by various methods (Heck 1981; Biacs 1989; Koch 1988), of which the Lenzmann–Heck's method will be specified in Sect. 7.3.2.

7.3 Methods of Verification of Altimetric Compatibility

7.3.1 Method of Height Congruency (Full Pre-information)

The height congruency of network points, vertically determined at epochs t and t', can be illustrated and presented in Fig. 7.1. The rectilinear connection of network points (in real: leveling connection) creates spatially dislocated area (generally for k points, a polyhedron is formed of triangular elements (plates) at both epochs), whose relative positions A, B illustrate different situations of altimetric compatibility of points 1, 2, 3.

If heights of points 1, 2, 3 at t and their positions 1', 2', 3' at t' creates stochastically congruent vertical geometry (practically the same area)—Fig. 7.1a at t and t'—it is a situation with a compatible vertical geodetic control.

If the points have not only stochastically identical heights at t and t' (heights will be significantly different at least on one point, e.g., point 3), points will create vertically non-congruent geometry of both areas—Fig. 7.1b—and in that case, it is a situation with an incompatible vertical geodetic control.

It is obvious that of the above options of the state of vertical geodetic control in practice, we expect or strive always to ensure the case A, i.e., a stochastic congruency of heights, in which only stochastic, insignificant influences mainly from measurements are always reflected.

Fig. 7.1 Height congruency
in the structure of 3 points;
A/acceptable congruency;
B/violated congruency (on the
point 3)

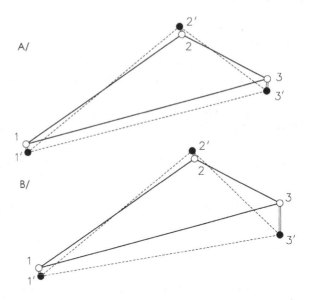

This phenomenon that has heights **h** and **h′** will never completely satisfy the identity **h** = **h′** from repeated measurements, but there will always be **h** ≠ **h′** and thus nonzero height differences (height discrepancies):

$$\mathbf{V_h} = d\mathbf{h} = \mathbf{h} - \mathbf{h'} \tag{7.2}$$

will always be on points, will allow us to determine which $d\mathbf{h}$ (on which point) is statistically significant and which is not significant, by assessment of values of $d\mathbf{h}$. Statistically significant values of $d\mathbf{h}$ identify those height points, which can be (with the risk α) classified as significantly changed points in height, i.e., points that are incompatible in height.

Formally, the method of height congruency has an analogous character as the method of planimetric congruency (Sect. Method of Planimetric Congruency) with the difference that in the present case, it is a congruency of one-dimensional variable $d\mathbf{h}$. The adjustment of vertical network at t and t' according to the Gauss–Markov singular model (GMM with an incomplete rank) is assumed.

For the application of this method, additional variables for test procedure are determined from values defined in Sect. 7.2.1, namely:

$d\mathbf{h} \equiv \mathbf{V_h} = \hat{\mathbf{h}}' - \hat{\mathbf{h}}$	height differences (discrepancies) at individual verified (datum) points,
$\mathbf{Q_{V_h}} = \mathbf{Q_{\hat{h}}} + \mathbf{Q_{\hat{h}'}}$	a cofactor matrix of height discrepancies,
$\mathbf{\Omega} = \mathbf{V_L^T Q_L^{-1} V_L}$	quadratic form of corrections $\mathbf{V_L}$ for measured differences in elevations $\mathbf{L} \equiv \Delta\mathbf{h}$,
$\Omega' = \mathbf{V_{L'}^T Q_{L'}^{-1} V_{L'}}$	quadratic form of corrections $\mathbf{V_{L'}}$ of measured differences in elevations $\mathbf{L}' \equiv \Delta\mathbf{h}'$,

$\mathbf{R} = \mathbf{V_h^T Q_{V_h}^{-1} V_h}$ quadratic form of height discrepancies $\mathbf{V_h}$ in the whole geodetic control,

$\mathbf{R}_i = \mathbf{V_{h_i}^T Q_{Vh_i}^{-1} V_{h_i}} = \frac{\mathbf{V_{h_i}^2}}{q_{Vh_i}}$ quadratic form of height discrepancy on the point i (simplified solutions of $\mathbf{R}_i$), where q_{Vh_i} is a cofactor of $\mathbf{V}_{h_i}$ discrepancy,

$\bar{s}_0^2 = \frac{\Omega + \Omega'}{(n-k+d)+(n'-k'+d)}$ the common a posteriori unit variance from both measurements and adjustments of height points at t and t' where d is a datum defect of the network at t and t'.

For a given case, the null hypothesis $H_0:E(\mathbf{V_h}) = 0$, which assumes $\mathbf{V_h}$ on points that have stochastic size, is formulated. The alternative hypothesis $H_a:E(\mathbf{V_h})$ / $= 0$ indicates that some discrepancies $\mathbf{V_h}$ represent significant value, indicating that the corresponding point is incompatible.

The used test variable (for global test, i.e., the whole vertical geodetic control) has the F-distribution (Niemeier 1980; Jakub 2001):

$$T = \frac{\mathbf{R}}{f_1 \bar{s}_0^2} \sim F(f_1, f_2) \tag{7.3}$$

with degrees of freedom, $f = k$, $f_2 = (n - k + 1) + (n' - k' + 1)$ [for the Gauss–Markov model with full rank $f_2 = (n - k) + (n' - k')$] and for the level of significance α (6.29), the critical value $((1 - \alpha)$ - quantile) of the F-distribution for f_1, f_2 will be defined as follows:

$$F_\alpha = F(\alpha; f_1, f_2). \tag{7.4}$$

If the comparison of T and F_α (the global test) gives:

$$T < F_\alpha,$$

H_0 is not rejected, and we can be confident about non-violating the height congruency in the structure of points with the risk α (i.e., virtually), therefore, convinced that all tested points have "changed" in height only stochastically during the period $t' - t$, i.e., all points are compatible in height; in the case of:

$$T \geq F_\alpha,$$

H_0 is rejected and contrary conclusions are adopted. In this case, at least one point that is incompatible and thereby also creates the incompatible character of the whole verified geodetic control of height points can be found in the set of verified points.

In this case, it is necessary to determine by a suitable procedure, which specific point is that incompatible point that causes the situation $T \geq F_\alpha$, from tested points P_1 to P_k.

The use of localization, point test [with modification of (7.3)], by which the $\mathbf{R}_i$ is examined for each point and thus also its compatibility in height, is one of the procedures.

Of other practical algorithms, it is also possible to use the global test in the cyclic realization of testing, which will be specified from now on.

By using this test, it can be determined by which portion individual points PH_i of vertical geodetic control contribute to the total value of $\mathbf{R}$ and which point contributes by the largest portion to the creation of the value $\mathbf{R}$. The point with the largest portion $\mathbf{R}_{i(max)}$ in the value of $\mathbf{R}$ will be an incompatible point that is searched (Pelzer 1971; Heck 1985; Biacs 1989), which also caused the incompatibility of the whole geodetic control.

Therefore, the problem to decompose $\mathbf{R}$ into its components $\mathbf{R}_i$ for every one height point arises for the realization of this procedure. The algorithm of decomposition of $\mathbf{R}$ (S-transformation of the vector $d\mathbf{h}$, van Mierlo 1980) is known (Illner 1983; Pelzer 1971; Caspary 1987), and we obtain values of $\mathbf{R}_i$ arranged directly in descending order beginning with the maximum value of $\mathbf{R}_{i(max)}$ when used with a suitable computing program.

The overall process of test procedure decomposed into test cycles is as follows.

0. Test Cycle

The whole vertical geodetic control is verified using the statistics (7.3) and the critical value of its distribution (7.4). If:

$$T < F_\alpha,$$

The geodetic control, i.e., all of its height points, is compatible and usable also for the current vertical geodetic activities, thereby the task, i.e., the examination of points for altimetric compatibility, ends.

In the case of:

$$T \geq F_\alpha,$$

the vertical geodetic control is incompatible; it is necessary to identify point(s) that caused incompatibility. For this purpose, it is necessary to determine the quadratic form of height discrepancies $\mathbf{R}$:

$$\mathbf{R} = \mathbf{V}_\mathbf{h}^\mathsf{T} \mathbf{Q}_{\mathbf{V}_\mathbf{h}}^{-1} \mathbf{V}_\mathbf{h} \tag{7.5}$$

and portions of $\mathbf{R}$ on individual points, i.e., values:

$$\mathbf{R}_1, \mathbf{R}_2, \ldots, \mathbf{R}_i, \ldots, \mathbf{R}_k,$$

that will be defined in a simplified expression as follows:

$$\mathbf{R}_i = \mathbf{V}_{\mathbf{h}_i}^\mathsf{T} \mathbf{Q}_{\mathbf{V}\mathbf{h}i}^{-1} \mathbf{V}_{\mathbf{h}_i} = \frac{\mathbf{V}_{\mathbf{h}_i}^2}{q_{\mathbf{V}\mathbf{h}i}}, \tag{7.6}$$

and which will identify the specific points incompatible in height within testing. The size of $\mathbf{R}_i$ indicates by how large proportion the individual points contribute to the creation of $\mathbf{R}$, while points with the highest $\mathbf{R}_i$ will be "responsible" for the incompatibility of the corresponding vertical geodetic control.

For example, let the results of $\mathbf{R}_i$ belong to individual points HL_1, HL_2, HL_3, ..., HL_k as follows:

1. **Test Cycle**

The maximum value of $\mathbf{R}_{1(max)}$ that belongs to the point HL_4, which will, therefore, be assumed that it caused the incompatibility of geodetic control, is taken from Table 7.1. The $\mathbf{R}$ is decreased by the value of $\mathbf{R}_1$, and the following test statistic is created for $\mathbf{R} - \mathbf{R}_1$:

$$T(1) = \frac{\mathbf{R} - \mathbf{R}_1}{f_1(1)\bar{s}_0^2} \sim F(f_1(1), f_2(1)), \qquad (7.7)$$

by which the testing of other points of geodetic control (except HL_4) is realized.

Also, the critical value of the F-distribution for α, $f_1(1) = k - 1$ and $f_2(1) = n_1 - (k - 1)$, where n_1 is the number of levelled differences in elevation in a vertical network after the elimination of point HL_4, is defined as follows:

$$F_\alpha(1) = (\alpha; k - 1, n_1 - (k - 1)). \qquad (7.8)$$

The global testing for the geodetic control of $k - 1$ height points (reduced by the removed point HL_4) is carried out with values of (7.7) and (7.8), with the result that also the original number of n leveling differences in elevation between points has changed to n_1, according to the particular situation. If:

$$T(1) < F_\alpha(1),$$

all of $k - 1$ tested points (HL_1, HL_2, HL_3, ..., HL_k) are compatible in height, only the point not taken into the testing—HL_4 with the maximum proportion of $\mathbf{R}_1$ in the value $\mathbf{R}$ is an incompatible point, thereby the end of testing takes place, and compatibility characteristics of points in a vertical structure are verified. If:

$$T(1) \geq F_\alpha(1),$$

geodetic control of verified $k - 1$ points is still incompatible and we assume that this incompatibility is caused by the point HL_3 with the second largest proportion of $\mathbf{R}_2$

Table 7.1 Results of R for individual HL points	The size ranking of $\mathbf{R}_i$	R_1	R_2	R_3	R_4	R_k
		max				min
	For height point	HL_4	HL_3	HL_1	HL_k	HL_2

of the value **R**. Therefore, it is necessary to proceed to the next cycle of testing, from which also the point HL_3 is removed and the global test is realized for the remaining points HL_1, HL_2, ..., HL_k.

2. **Test Cycle**

The second largest proportion of $\mathbf{R}_2$ in the **R** corresponding to the point HL_3 is taken from Table 7.1, it is also subtracted from **R** (together with $\mathbf{R}_1$), and the test statistic is thus determined according to:

$$T(2) = \frac{\mathbf{R} - (\mathbf{R}_1 + \mathbf{R}_2)}{f_1(2)\bar{s}_0^2} \sim F(f_1(2), f_2(2)), \tag{7.9}$$

as well as the critical value of the F-distribution for α, $f_1(2) = k - 2$ and $f_2(2) = n_2 - (k - 2)$:

$$F_\alpha(2) = F(\alpha; k - 2, n_2 - (k - 2)), \tag{7.10}$$

where n_2 is the number of leveling differences in elevation in the remaining network structure after the deletion of points HL_4 and HL_3. If the following situation arises from the comparison of (7.9) and (7.10):

$$T(2) < F_\alpha(2),$$

all of the $k - 2$ currently tested points are compatible; thus, even the point HL_3 is an incompatible point and a total of 2 incompatible points HL_4 and HL_3 are in the verified geodetic control. Thereby, the verification of height points ends and the remaining points HL_1, HL_2, ..., HL_k can be considered as point compatible in height; in the case of:

$$T(2) \geq F_\alpha(2),$$

geodetic control of verified $k - 2$ points is still incompatible, and we assume that it is caused by the point HL_1 with the third largest portion of $\mathbf{R}_3$ in the value **R**, and therefore, we have to proceed to another (3rd) cycle of global testing by such procedure as outlined in the 1st and 2nd cycle.

In general, the testing in cycles by the procedure outlined continues until the jth cycle, in which the geodetic control of not removed points is proven as compatible.

Example No. 5 Let us have a local level network (LLN) of 6 points HL_1, ..., HL_6 of national-level networks of various orders (of superior altimetric networks) (Fig. 7.2) established in a certain area in the past (epoch t). These points form a LLN that was surveyed and adjusted on the level of precise leveling.

The full documentation of its adjustment at this epoch t is available.

At present (epoch t'), it is necessary to use height points of this local network for current surveying tasks, and it is therefore necessary to verify their altimetric

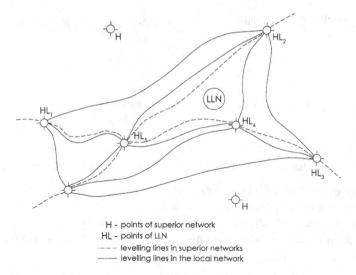

H - points of superior network
HL - points of LLN
‑‑‑‑ levelling lines in superior networks
——— levelling lines in the local network

Fig. 7.2 Local-level network (LLN) at epochs t and t'

compatibility. For this purpose, the network was surveyed and adjusted at t' (Gauss–Markov singular model: adjustment of the free network), and thus, the complete documentation on the current state of the network and its points is also available.

The examination of altimetric compatibility of leveling points in such a case (i.e., in a given pre-information situation) should be realized by the method of height congruency, the use of which is as follows:

The following data and variables are used from the epoch t (Table 7.2):

the number of leveling differences in elevation $n = 11$,
the number of leveling points—heights $k = 6$,
6×6 cofactor matrix of adjusted heights $\mathbf{Q}_{\hat{h}}$ (not specified),
quadratic form of corrections $\mathbf{V}$ of differences in elevation $\mathbf{L}{:}\mathbf{\Omega} = 0.000029705$,
a posteriori unit variance $s_0^2 = 0.000004951$.

Table 7.2 Adjusted heights of HL points—epoch t

Point	Adjusted heights $\hat{h}$ (m)
HL_1	527.2135
HL_2	506.5537
HL_3	535.3791
HL_4	561.0766
HL_5	550.3038
HL_6	544.1953

Table 7.3 Adjusted heights of HL points—epoch t'

Point	Adjusted heights $\hat{h}'$ (m)
HL_1	527.2152
HL_2	506.5523
HL_3	535.3784
HL_4	561.0931
HL_5	550.3050
HL_6	544.1966

The following data and variables are used from the epoch t' (Table 7.3):

the number of leveling differences in elevation $n' = 11$,
the number of leveling points—heights $k' = 6$,
6×6 cofactor matrix of adjusted heights $\mathbf{Q}_{\hat{h}'}$ (not specified),
quadratic form of corrections $\mathbf{V}'$ of differences in elevation $\mathbf{L}':\Omega' = 0.00032491$,
a posteriori unit variance $s_0'^2 = 0.000005415$.

And additional variables for testing are determined from these values (Table 7.4):

a cofactor matrix of height discrepancies $\mathbf{Q}_{dh} = \mathbf{Q}_h + \mathbf{Q}_{\hat{h}'}$ (not specified),
a quadratic form of height discrepancies $\mathbf{R} = 0.000192475$,
the common unit variance $\bar{s}_0^2 = 0.000005183$, and
quadratic forms of discrepancies on individual points HL_i according to the size of $\mathbf{R}_i$.

The following is determined in each cycle of the test algorithm, regarding Sect. 7.3.1:

0. **Cycle**

The statistics T and $(1 - \alpha)$ - quantile of the F-distribution for the global test of geodetic control with all points for $\alpha = 0.05$ and $f_1 = 6, f_2 = 2(11 - (6 - 1)) = 12$ will be defined as follows:

Table 7.4 Height discrepancies of HL points

Point	Height discrepancies dh (m)
HL_1	0.0017
HL_2	−0.0014
HL_3	−0.0007
HL_4	0.0115
HL_5	0.0012
HL_6	0.0023

$$T = \frac{\mathbf{R}}{f_1 \bar{s}_0^2} = \frac{0.000192475}{6 \times 0.000005183} = 6.1893,$$

$$F_\alpha = F(\alpha; f_1, f_2) = F(0.05; 6, 12) = 2.9961,$$

by using which we get:

$$T > F_\alpha.$$

Thus, the vertical geodetic control of 6 points is incompatible, and it is necessary to find the point(s) that have caused it. For this purpose, the required number of test cycles is realized, in which values of $\mathbf{R}_i$ resulting from a simplified decomposition of $\mathbf{R}$ corresponding to individual points are used (Table 7.5).

1. Cycle

The point HL_4 with the maximum value of $\mathbf{R}_i$, which could most likely cause the incompatibility of geodetic control, is excluded from the geodetic control of height points, resulting in $\mathbf{R}$ being decreased to value $\mathbf{R} - \mathbf{R}_1$. For statistics (7.9) and critical value (7.10), we get:

$$T(1) = \frac{\mathbf{R} - \mathbf{R}_1}{f_1 \bar{s}_0^2} = \frac{0.000031387}{5 \times 0.000005183} = 1.2112,$$

$$F_\alpha(1) = F(\alpha; f_1, f_2) = F(0.05; 6 - 1 = 5, (8 + 8) - (5 + 5) = 6) = 4.3837,$$

where the numbers $n = n' = 11$ of original measurements and $k = k' = 6$ points were decreased to $n = n' = 8$ and $k = k' = 5$ by the exclusion of point HL_4. Their comparison gives:

$$T(2) < F_\alpha(2),$$

and thus, all 5 tested points in the 1st cycle are compatible, but only point HL_4 is incompatible in height, resulting at the end of testing.

Therefore, as it follows from the process and results, all points of the vertical geodetic control with points HL_1, ..., HL_6, except for point HL_4, are compatible, and height discrepancies $d\mathbf{h}$ on these points have only stochastic sizes and can be

Table 7.5 Quadratic form of discrepancies

Point	R_i
HL_4	0.000161088
HL_6	0.000010503
HL_1	0.000005814
HL_2	0.000005537
HL_5	0.000004955
HL_3	0.000004669

used, with their heights $\hat{h}_1, \hat{h}_2, \hat{h}_3, \hat{h}_5, \hat{h}_6$, for the current geodetic activities realized at the t'.

7.3.2 Lenzmann–Heck's Method (Incomplete Pre-information)

This method, analogously to its application for the verification of planimetric compatibility (Sect. Lenzmann–Heck's Method), represents an identical procedure also in the present case when in use for the verification of altimetric compatibility of points, both in the preparation of data for testing and in the actual testing process with the difference that procedures are now related to one-dimensional variables.

Use of the method can be demonstrated on the following model.

Geodetic activities should be realized in a certain area. Leveling points H from level networks of various order with known heights h_i in the system of Bpv (epoch t), which are available, are situated in the wider surroundings of a given area as well as in it. From the set of height points nearest to the area of interest, $k = 6$, points HL_i, $i = 1, 2, \ldots, 6$ (Fig. 7.3), were selected, while these points should be used as connecting (datum) points for height problems in the considered area, for example, for the establishment of new local height points U. It is, therefore, necessary to verify points HL in terms of their altimetric compatibility, so that possible defective height points can be excluded from them and thus that any height problem can be solved with the same quality when connected to any points from the set of HL_i.

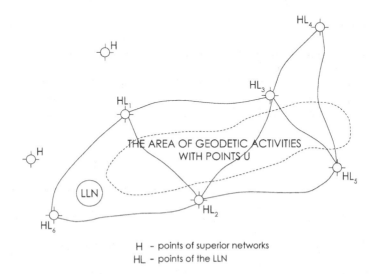

H – points of superior networks
HL – points of the LLN

Fig. 7.3 Local-level network (LLN) from the epoch t

For this purpose, a LLN of points HL_i (Fig. 7.3) is established. In accordance with the fundamental principle of the verification of altimetric compatibility, it is also necessary to perform their current survey (epoch t') and determination of heights of these points HL_i.

In order that results of the current survey of the LLN with points U, HL, and its adjustment were influenced only by measurements, it is appropriate to adopt a local vertical system S(LOC) for the adjustment of current measurements. In this system, local heights of points $\hat{h}_i^L$ are obtained from the survey and adjustment of LLN (Gauss–Markov singular model is suitable, i.e., the adjustment of the vertical net-work by GMM with incomplete rank). These heights are transformed to heights h_i^t in the S(Bpv) by translational (isometric) transformation (Böhm et al. 1981). Height differences dh_i ∴ discrepancies $\mathbf{V}_{h_i}$ obtained from the comparison of h_i and h_i^t form the basis for the assessment of the compatibility of corresponding points HL_i.

In terms of the verification procedure, it is, therefore, necessary to prepare the following variables:

from the epoch t: $\mathbf{h}_i$database heights in the S(Bpv) of points HL_i and

from the epoch t': $\hat{\mathbf{h}}_i^L$local heights of points HL_i in the selected S(LOC).

Subsequently, it is necessary to perform also 2D translation transformation, in which the deterministic model of transformation equations is as follows:

$$\mathbf{h}^t = \mathbf{h}^L + \hat{\mathbf{t}}\mathbf{p}, \tag{7.11}$$

and the corresponding statistical model is as follows:

$$h_i^t = \hat{h}_i^L + e\hat{\mathbf{t}}\mathbf{p}, \quad i = 1, 2, \ldots, 6, \quad e^T = [111111], \tag{7.12}$$

where

$$\hat{\mathbf{t}}\mathbf{p} = \sum_1^6 tp_i/6 \tag{7.13}$$

is the estimation of the transformation parameter from its realizations:

$$tp_i = h_i^t - \hat{h}_i^L. \tag{7.14}$$

Therefore, S(Bpv)—heights of points HL_i obtained by transformation are as follows:

$$\mathbf{h}^t = \hat{\mathbf{h}}^L + e\hat{\mathbf{t}}\mathbf{p}, \quad e = [111111]^T, \tag{7.15}$$

while their accuracy is described by the covariance matrix:

$$\Sigma_{\mathbf{h}^t} = \Sigma_{\hat{\mathbf{h}}} + \Sigma_{\hat{\mathbf{tp}}} \tag{7.16}$$

in which the covariance matrix $\Sigma_{\hat{\mathbf{h}}} = s_0^2 \mathbf{Q}_{\hat{\mathbf{h}}^L}$ of estimates of local heights $\hat{h}_i^L$ is obtained from the adjustment of vertical network at the time t', and the covariance matrix of transformation parameter $\hat{\mathbf{tp}}$ is determined according to:

$$\Sigma_{\mathbf{tp}} = \text{diag}(s_{\mathbf{tp}}^2 e), \tag{7.17}$$

where

$$e^T = [111111]$$

is a unit vector, and the variance of estimation of transformation parameter is defined as follows:

$$s_{\mathbf{tp}}^2 = \frac{\Sigma(\hat{\mathbf{tp}} - tp_i)^2}{k(k-1)}. \tag{7.18}$$

Each point HL$_i$ can, therefore, be assessed by comparing its height h_i (from the epoch t) and h_i^t (from the epoch t'), which should be stochastically identical, i.e., their differences (discrepancies):

$$dh_i = h_i - h_i^t \tag{7.19}$$

should be the values close to zero. Therefore, if the corresponding dh_i is a non-random, numerically significant value for some point HL$_i$, it will indicate that such a point HL$_i$ is defective in terms of its altimetric compatibility.

The accuracy of the k-dimensional vector of height discrepancies $d\mathbf{h}$ is described by the following $k \times k$ covariance matrix:

$$\Sigma_{d\mathbf{h}} = \Sigma_{\mathbf{h}} + \Sigma_{\mathbf{h}^t}, \tag{7.20}$$

where $\Sigma_{\mathbf{h}}$ is usually only diagonal matrix with variances s_h^2 of heights h (generally with approximate values from the epoch t) of k points and $\Sigma_{\mathbf{h}^t}$ is given by (7.16).

A random variable (Heck 1985) for every height discrepancy will be the localization test statistic in the present case for dh_i, analogous to the expression (6.58) for 2D variables:

$$T_i = \frac{\mathbf{R}_i}{r\hat{s}_{0_i}^2} \sim F(f_1, f_2), \tag{7.21}$$

where degrees of freedom of the F-distribution are as follows:

$$f_1 = d = 1,$$
$$f_2 = 2k - tp - d = 10,$$

(7.22)

with values (for the example No. 6):
$k = 6$ the number of height points in the network,
$tp = 1$ the number of transformation parameters, and
$d = 1$ the coordinate dimension of points;
and other variables (with known meaning—Sect. Lenzmann–Heck's Method):

$$\hat{s}_{0_i}^2 = \frac{R - R_i}{k - tp - d},$$

(7.23)

$$R = d\mathbf{h}^T Q_{d\mathbf{h}}^{-1} d\mathbf{h},$$

(7.24)

$$R_i = d\mathbf{h}_i^T Q_{d\mathbf{h}_i}^{-1} d\mathbf{h}_i = \frac{dh_i^2}{q_{dh_i}}.$$

(7.25)

The critical value of the F-distribution in the selection of the level of significance α (6.29), when $\alpha_0 = 0.01$ is used for each identification testing (for each height discrepancy), is as follows:

$$F(\alpha_i) = F(\alpha_0; f_1, f_2).$$

(7.26)

Also in the present case, the null and alternative hypotheses are formulated for testing:

$$H_{0_i}: dh_i = 0, \ H_{a_i}: dh_i \neq 0,$$

(7.27)

expressing the assumption that dh_i will be a statistically insignificant value (H_{0_i}), respectively, that dh_i represents a statistically significant value (H_{a_i}).

Subsequently, we get either the following relation from the comparison of T_i and $F(\alpha_i)$ for individual points:

$$T_i < F(\alpha_i),$$

when H_{0_i} is not rejected, i.e., the relevant dh_i can be considered as stochastic value, and the corresponding point is, therefore, compatible in height, usable also for the current survey, or:

$$T_i \geq F(\alpha_i),$$

when H_{0_i} is rejected, i.e., the relevant indicator dh_i is considered as statistically significant value. Thus, the relevant point can be considered as incompatible in height, with the risk α_0 of incorrect rejection of H_{0_i}.

Example No. 6 Consider a situation for the verification of altimetric compatibility for the model in the Fig. 7.3. Altitudes h_i of points HL_i, $i = 1, 2, ..., 6$, in the system S(Bpv) are given (epoch t) from the established level networks, which should be currently (epoch t') verified by the test according to Sect. 7.3.2.

According to (6.14), realizations dp_i of the transformation parameter in the used 1D translation transformation $\hat{h}_i^{\prime L} \Rightarrow h_i^{tBpv}$ were derived from given heights h_i of points HL_i and their local heights $\hat{h}_i^{\prime L}$ obtained from the current survey and adjustment in the local vertical system. Values of relevant variables are listed in Table 7.6.

The estimation of transformation parameter was determined on the basis of LSM adjustment of direct measurements tp_i:

$$\hat{\mathbf{tp}} = 2.27495[\mathrm{m}],$$

and by means of it, transformed heights h^t in the system S(Bpv) (Table 7.6) were derived according to (7.14). Height discrepancies (indicators) $dh_i \equiv V_{h_i} = h_i - h_i^t$ (Table 7.7) were examined for testing: values $\mathbf{R}_i = dh_i^2$ according to (7.25) with $\forall q_{dh} = 1$ (Table 7.8), the value of $\mathbf{R}$ according (7.24):

$$\mathbf{R} = \Sigma dh^2 = 0.000025227,$$

and values of $\hat{s}_{0_i}^2$ according to (7.23) (Table 7.9).

Table 7.6 Values of required variables

Point	$h\varphi$ (m)	$\hat{h}^{\prime L}$ (m)	tp (m)	h^t (m)
HL_1	217.2747	215.0000	2.2748	217.2750
HL_2	217.1567	214.8855	2.2761	217.1605
HL_3	217.0472	214.7728	2.2738	217.0478
HL_4	217.0753	214.8025	2.2755	217.0775
HL_5	217.8699	215.5972	2.2742	217.8721
HL_6	217.6591	215.3831	2.2753	217.6580

Table 7.7 Height discrepancies of HL points

Point	dh (m)
HL_1	−0.00028
HL_2	−0.00374
HL_3	−0.00051
HL_4	−0.00214
HL_5	−0.00228
HL_6	0.00106

Table 7.8 Values of R_i for individual HL points

Point	R_i
HL_1	7.84e-8
HL_2	0.000014
HL_3	2.601e-7
HL_4	4.5796e-6
HL_5	5.1984e-6
HL_6	1.1236e-6

Table 7.9 Values of $\hat{s}_{0_i}^2$ for individual HL points

Point	$\hat{s}_{0_i}^2$
HL_1	2.5149e-6
HL_2	1.1240e-6
HL_3	2.4968e-6
HL_4	2.0648e-6
HL_5	2.0029e-6
HL_6	2.4104e-6

Realizations of statistics T_i (7.21) for individual points HL_i after substituting these values to corresponding variables are shown in Table (7.10), and its critical value according to (7.26) is as follows:

$$F_{\alpha_0} = (0.01; 1, \; 10) = 10.044.$$

The following hypotheses are adopted for testing:

$$H_{0_i}:dh_i = 0, \; H_{a_i}:dh_i \neq 0,$$

with the usual meaning and interpretations, and they are assessed based on the comparison of corresponding T_i and F_{α_0} for individual points. From the comparison of T_i and F_{α_0} for individual points, we get:

Table 7.10 Values of statistics T_i for individual HL points

Point	T_i
HL_1	0.0312
HL_2	12.4444
HL_3	0.1042
HL_4	2.2179
HL_5	2.5954
HL_6	0.4661

$$\begin{aligned}
\text{HL}_1 \quad & T < F_{\alpha_0} \\
\text{HL}_2 \quad & T > F_{\alpha_0} \\
\text{HL}_3 \quad & T < F_{\alpha_0} \\
\text{HL}_4 \quad & T < F_{\alpha_0} \\
\text{HL}_5 \quad & T < F_{\alpha_0} \\
\text{HL}_6 \quad & T < F_{\alpha_0},
\end{aligned}$$

and therefore, the following conclusions result for the assessment of the altimetric compatibility of points HL_i, $i = 1, \ldots, 6$:

- H_{0_i} are not rejected for points HL_1, HL_3, HL_4, HL_5, HL_6, which can be considered as points compatible in height at the epoch t, suitable also for the current use, based on their height discrepancies $d\mathbf{h}$ and their semantic interpretation on the level of significance α_0;
- H_{0_i} is rejected for point HL_2, i.e., this point can be considered as incompatible in height and unusable for surveying practice at the epoch t' on the basis of statistical assessment of the corresponding discrepancy dh_2.

Chapter 8
The Verification of Compatibility of Spatial Points

8.1 In General

The verification of the compatibility of points of spatial networks forms a very important component of completion and extension of networks by new points required for accurate geodetic activities. Therefore, it is necessary to have a quality geodetic control. A quality geodetic control can be achieved not only by new and quality measurement of new points, but also fine connecting points of which a new geodetic control is being established must be available.

The verification itself has already been described in chapters on the verification of planimetric and height points, where it is already stated that the compatibility can be verified using the $d\mathbf{L}$ or $d\mathbf{C}$ indicators. For spatial networks, it is more advantageous (or more transparent) to verify the compatibility using $d\mathbf{C}$ indicators.

8.2 The Verification of Spatial Compatibility by DC Indicators

The verification of compatibility using indicators $d\mathbf{C}$ is more advantageous especially for the exactness of characterization and representation of both components of a point, namely the physical stability of a survey mark and the coordinate stability of a point of the spatial network. $d\mathbf{C}$ indicators directly indicate the condition of points of spatial geodetic control, whether they are compatible or incompatible points. $d\mathbf{L}$ indicators, in contrast to $d\mathbf{C}$ indicators, do not provide direct information on the compatibility of point/points, because measured variables are in a functional relationship with determined coordinates and cannot provide direct characteristics of the examined point. $d\mathbf{L}$ indicators should be used for the verification of compatibility of one or a couple of points, and the use of $d\mathbf{L}$ indicators is appropriate for network structures.

© The Author(s) 2016
G. Weiss et al., *Survey Control Points*,
SpringerBriefs in Geography, DOI 10.1007/978-3-319-28457-6_8

8.3 The Verification of Spatial Compatibility with Pre-information

It is necessary to have two surveys of the network at two epochs in order to be able to examine whether the spatial network is compatible. Since the compatibility is being examined, the current survey of the network at the epoch t' must have a complete documentation of the survey and processing of geodetic control. The verification of compatibility can be realized in two different situations: either if all the necessary data from the previous epoch t are available (i.e., the complete documentation of survey and processing of geodetic control), or if the documentation is incomplete (many times, only coordinates of points of the corresponding geodetic control are available).

According to Sect. 3.5, the documentation and following data from the epoch t are necessary for the verification of compatibility of geodetic control:

$$\mathbf{C}, \boldsymbol{\Sigma}_\mathbf{C}, s_0^2, \mathbf{L}, \mathbf{Q}_\mathbf{L}, \mathbf{V}, \ldots \tag{8.1}$$

from the survey and processing of network.

Analogously to Sect. 6.2.1.1, points PL_i belonging to the superior network and new points B_i, which are used to densify the geodetic control and will be used for various geodetic activities and calculations, are illustrated in Fig. 8.1. Using the

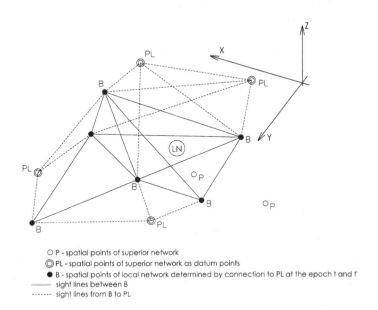

O P - spatial points of superior network
◎ PL - spatial points of superior network as datum points
● B - spatial points of local network determined by connection to PL at the epoch t and t'
———— sight lines between B
------- sight lines from B to PL

Fig. 8.1 A local spatial network of points B, PL, surveyed at epochs t and t'

same conditions and methods as at the epoch t, the same range of data and documentation is obtained from the current epoch t':

$$\mathbf{C}', \mathbf{\Sigma}_{\mathbf{C}'}, s_0'^2, \mathbf{L}', \mathbf{Q}_{\mathbf{L}'}, \mathbf{V}', \ldots \tag{8.2}$$

From the complete documentation of both epochs and other variables derived from them, in particular dC indicators of compatibility and their accuracies are verified by a suitable statistical hypothesis testing, whether the geodetic control of spatial network is compatible or incompatible, or which points of the network are incompatible (and, therefore, which points should not be used for further geodetic activities and processing).

In contrast to the verification of compatibility of planimetric geodetic control, the necessary data are also available (8.1), (8.2), and also the indicators of spatial network dC will have the form corresponding to the following structure:

$$dC = C' - C = \begin{pmatrix} \vdots \\ dC_i \\ \vdots \end{pmatrix} = \begin{pmatrix} \vdots \\ C_i' \\ \vdots \end{pmatrix} - \begin{pmatrix} \vdots \\ C_i \\ \vdots \end{pmatrix} = \begin{pmatrix} \vdots \\ dX_i \\ dY_i \\ dZ_i \\ \vdots \end{pmatrix} = \begin{pmatrix} \vdots \\ X_i' - X_i \\ Y_i' - Y_i \\ Z_i' - Z_i \\ \vdots \end{pmatrix} \tag{8.3}$$

At first, it must be examined if the both epochs are at the same level of accuracy (Sect. 8.3.1). Subsequently, it is possible to realize a global test of the spatial network (Sect. 8.3.1), by which it is determined (based on the vector dC) if any incompatible points are located in the network. Afterward, it is necessary to perform the localization test (Sect. 8.3.3) by which the points of incompatibility are localized (or if all points are compatible) based on the corresponding components $dC_i = (dX', dY', dZ')$ of the point C_i.

The new processing of network is realized without the use of an incompatible point, and the subsequent spatial network should already be compatible if the contrary is not confirmed.

8.3.1 Fisher's Compatibility Test of Two Files with Different Accuracy

In case of a separate adjustment of deformation network, individual epochs or their measured data are adjusted separately. Therefore, two sets of data with an unequal accuracy are obtained. The question is, whether they can be further processed together. This can be assessed through the unit a posteriori factors s_0^2 as the best mathematical expressions for the accuracy of data samples obtained from a processing of network structures.

If the initial processing of networks is correct and no systematic errors are present in data, then the ratio of two compared variances should be close to 1. H_0 of a test is designed to compare the ratios of variances for two selections as follows (Böhm et al. 1981):

One-tailed test
Two-tailed test

The null hypothesis:

$$H_0 : \frac{s^2}{s'^2} = 1 \text{ resp. } s^2 = s'^2, \quad H_0 : \frac{s^2}{s'^2} = 1 \quad \text{resp. } s^2 = s'^2 \tag{8.4}$$

The alternative hypothesis:

$$H_A : \frac{s^2}{s'^2} < 1 \text{ or } s^2 < s'^2, \quad H_A : \frac{s^2}{s'^2} \neq 1 \text{ or } s^2 \neq s'^2,$$

$$H_A : \frac{s^2}{s'^2} > 1 \text{ or } s^2 > s'^2. \tag{8.5}$$

The test statistic, which is used to determine the acceptance/rejection of the null hypothesis H_0, is given as:

$$T = \frac{s^2}{s'^2}, (s^2 \geq s'^2); \quad T \approx F_{\alpha/2}(f, f') \tag{8.6}$$

where $f = (n - k)$ and $f' = (n' - k')$. The null hypothesis H_0 is rejected if it fulfills the following statement:

$$T > F_\alpha, \quad T > F_{\alpha/2} \tag{8.7}$$

Therefore, there is no possible common processing of these sets of observations in one mathematical model.

8.3.2   Global Test of Compatibility of Spatial Network

A significant stability or instability of network points is examined by a global compatibility test, for which the null hypothesis is expressed as (Weiss and Jakub 2007):

$$H_0 : dC = C' - C = 0, \tag{8.8}$$

expressing the assumption that network points remain stable over time $t' - t$, compared to the alternative hypothesis:

$$H_A : dC = C' - C \neq 0. \tag{8.9}$$

This H_0 is confronted with the reality and is either accepted on a certain selected probability level (as conforming to reality) or is rejected. These possible standpoints on a stability of points are accepted by comparing the test statistic for compatibility tests:

$$T = \frac{\Omega_{dC}}{s_0^2 f} \sim F_\alpha(f_1, f_2), \tag{8.10}$$

where $\Omega_{dC} = dC^T \cdot Q_{dC}^{-1} \cdot dC$ is a quadratic form of coordinate deviations and $f_1 = k$, $f_2 = n - k$.

In case of a separate adjustment of spatial network, the cofactor matrix of deformation vector is calculated by:

$$Q_{dC} = Q_{C'} + Q_C, \tag{8.11}$$

and in case of bivariate processing:

$$Q_{dC} = Q_C + Q_{C'} - 2Q_{C''}, \tag{8.12}$$

where s_0^2 is an a posteriori variance factor obtained from the adjustment. In case of separate adjustment, it should be determined as an average value of a posteriori variance factors from both adjustments:

$$\bar{s}_0^2 = \frac{v^T \cdot Q_L^{-1} \cdot v + v'^T \cdot Q_L'^{-1} \cdot v'}{(n - k) + (n' - k')}, \tag{8.13}$$

where $f_1 = k, f_2 = n - k$ are the degrees of freedom of the F-distribution of random variable T and its critical value for a level of significance $1 - \alpha : F_\alpha(1 - \alpha, f_1, f_2)$.

If $T < F_\alpha$, H_0 is not rejected, it is accepted, all determined points of the spatial network can be considered stable, without the effect of such deformation forces that would have changed their position, at the epoch $t' - t$. The network implementation is now compatible.

If $T > F_\alpha$, H_0 is rejected, then some of the points have significantly changed their position for the period $t' - t$, due to the effects of deformation forces (Sütti et al. 1997; Sabová and Jakub 2007).

8.3.3 *Local Test of Compatibility of Spatial Network*

For the localization of changed 3D points, a numerical value of R (applies to p points) is decomposed into its partial components R_i, related to the individual points of the network. Decomposition can be done by Weiss et al. 2007:

- the exact procedure requiring a specific algorithm of decomposition of R to values R_i corresponding to individual points of the network, in which values of all autocorrelation and intercorrelation ties of coordinate determination of points, as well as the influences of measurement and calculations, will be considered,
- the approximating procedure, in which its proportion in R_i will be determined for each point, by the value:

$$R_i = dC_i^T \cdot Q_{dC_i}^{-1} \cdot C_i, \tag{8.14}$$

expressed by relevant elements only from the main diagonal of the matrix Q_{dC}, which means ignoring the effect of correlation relations, which generally do not affect point values of R to such an extent, that the test result has been changed (3D stable point $\sim$ 3D changed point).

The localization test statistic T_i for individual points of spatial network is determined by:

$$T_i = \frac{R_i}{k_1 \bar{s}_0^2}, \quad i = 1, 2, \ldots, k_1. \tag{8.15}$$

The critical value for Fisher's distribution:

$$F_{\text{crit}} = F_\alpha(1 - \alpha, k_1, n - k). \tag{8.16}$$

Significant change in spatial position of point for the time period $t' - t$ is not proven by the test, if:

$$T_i \le F_{\text{crit}}. \tag{8.17}$$

Therefore, changes of point coordinates at t' against their values at t are not significant, and the point can be considered as stable for a given period. The instability of the point is expressed by:

$$T_i > F_{\text{crit}}, \tag{8.18}$$

a spatial change of relevant point can be accepted at the level of significance α as a result of deformation forces (Weiss et al. 2007).

8.3.4 Test by Confidence Ellipsoid

Similarly to describing the accuracy of determination of point position in a plane by standard ellipses, ellipsoids can be used for the graphical determination of deformations of a point in space. This type of ellipsoids is known as confidence ellipsoids because they express the significance of spatial changes.

The calculation of parameters of relative ellipsoids is based on a cofactor matrix of deformation vector Q_{dC}, which is subject to spectral decomposition for the determination of the main parameters of the ellipsoid.

This ellipsoid is constructed with certain probability at the point B at the epoch t. In general, 3 cases can occur in the determination of instability of point (Fig. 8.2) (Weiss et al. 2007; Sabová et al. 2007):

1. If the displacement vector $\Delta\Theta_B = \overline{^1B^2B} = \sqrt{\Delta X^2 + \Delta Y^2 + \Delta Z^2}$ of the point B between epochs t and t' will be located inside the ellipsoid, a standpoint may be accepted with the probability $1 - \alpha$, that coordinate changes $\Delta X, \Delta Y, \Delta Z$ of the point B at the epoch t, t' are not significant, i.e., position of the point ^2B in space relative to ^1B has a stochastic character, caused only by different random conditions and influences.

2. If the endpoint ^3B of the vector $\Delta\Theta_B = \overline{^1B^3B}$ is located on the surface of an ellipsoid, a suspicion of instability arises, but we cannot clearly decide whether a displacement occurs or the end position of a point ^2B is just the result of accumulated errors of measurement.

3. If the vector $\Delta\hat{\Theta}_B = \overline{^1B^4B}$ will permeate the confidence ellipsoid, i.e., its endpoint ^4B will be located outside of the ellipsoid of the point ^1B at the epoch t', then this indicates that the point ^4B has been significantly displaced due to the deformation forces in the relevant area over a period of t, t'.

Fig. 8.2 Displacements of the point ^1B up to the point ^4B

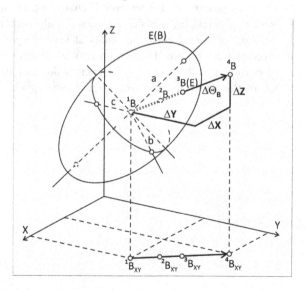

Assessment of significance of 3D changes of determined points in the spatial network between individual epochs is essentially also possible by the test of 3D compatibility of spatial realizations of the network. However, due to the different quality of positional and vertical determination of points by GNSS technology, it is better to prefer point tests.

8.3.5   Graphical Representation of Results

A variety of graphical representations of deformation phenomena are widely used for better illustration of processing results of the observed network (movements, and/or stability of its points). They are mainly used to display the progress of changes between a pair of epochs, or in a time interval of realization of several epochs and they give a better idea of changes of network structures on a global scale as well as detailed views on individual components.

Graphical representation of points movement can be achieved as clearly as possible by displacement vectors of points for a period of t and t', 3D visualization of surfaces and objects, their cross sections or isoline representation of the size and speed of changes in the space of deformation network for a certain period.

For vector representation of the displacements of different situations, their extreme cases are typical when:

- Vectors are chaotically distributed without any apparent trend in the surrounding area of the initial epoch. Their continuance progress is in different directions and with different sizes in the examined period, while some changes of points at epochs may also show significant values. However, they tend to be interpreted as the effect of local forces in the neighborhood of a point or incorrect measurement process for longer overall change of a point without a trend.
- Vector progress in certain directions with certain trends in strips, where deformation forces affect on points at several epochs. Partial vectors may be diverted from the trend at certain epochs, or eventually their size is inversely proportional to the size of other vectors due to local forces, but globally they follow this trend, they "return" to it at next epochs (Sabová et al. 2007).

References

Anděl J (1972) Matematická statistika. SNTL, Praha

Antonopoulos A (1985) Zur Formulierung und Überprüfung von Deformationsmodellen. Wissenschaftliche Arbeiten der Fachrichtung Vermessungswesen der Universität Hannover, Nr. 138, Universität Hannover

Baarda W (1968) A testing procedure for use in geodetic networks. Publications on Geodesy, New series, vol 2, No. 5, Technical University of Delft

Benning W (1985) Test von Ausreissern bei der Helmerttransformation. 110(5):207–209

Biacs ZF (1989) Estimation and hypothesis testing for deformation analysis in special purpose networks. UCSE Report No. 20032, University of Calgary

Bill R (1984) Eine Strategie zur Ausgleichung und Analyse von Verdichtungsnetzen. Verlag der Bayerischen Akademie der Wissenschaften in Kommission bei der C.H. Beck'schen Verlagsbuchhandlung, vol 295, München

Böhm J et al (1981) Vyšší geodézie. Díl 2, Czech Technical University in Prague, Praha

Boljen J (1986) Identitätsanalyse Helmert-transformierter Punkthaufen. Zeitschrift fur Vermessungswesen 111(11):490–500

Caspary W (1987) Concepts of network and deformation analysis. Monograph No. 11, School of Surveying, University of New South Wales, Kensington

Cimbálnik M (1978) Současné geodetické sítě z hlediska optimalizace. Výzk. Zpráva, Research Institute of Geodesy, Topography and Cartography (VÚGTK), Praha

Czaja J (1996) Aproksymacja wektorowego pola przemieszczeń oraz jego interpretacja geometryczna i fizyczna. Geodezja i kartografia, vol XX, No. 4, Warszawa

Delong B (1960) Zhodnocení Československé trigonometrické sítě I. řádu. Geodetický a kartografický sborník, vol 6. SNTL, Praha

Erker E (1997) Die Homogenisierung des österreichischen Festpunktfeldes im internationalen Rahmen. VGI – Österreichische Zeitschrift für Vermessung und Geoinformation 3(2):109–116

Fotiou A et al (1993) Adjustment, variance components estimation and testing with the affine and similarity transformations. Zeitschrift fur Vermessungswesen 118(10):494–503

Gründig L et al (1985) Detection and localization of geometrical movements. J Surveying Eng 111 (2):118–132

Hald A (1972) Statistical theory with engineering applications. Wiley, New York

Hanke K (1988) Eliminierung der nicht-signifikanten Parameter bei der Transformation zwischen ungleichartigen Koordinatensystemen. Österreichische Zeitschrift für Vermessungswesen und Photogrammetrie 76(4):432–439

Heck B (1985) Ein- und zweidimensionale Ausreissertests bei der ebenen Helmerttramsformation. Zeitschrift fur Vermessungswesen 110(9):461–471

Heck B (1981) Der Einfluss einzelner Beobachtungen auf das Ergebnis einer Ausgleichung und die Suche nach Ausreissern in den Beobachtungen. Allgemeine Vermessungs-Nachrichten 88 (1):17–34

© The Author(s) 2016
G. Weiss et al., *Survey Control Points*,
SpringerBriefs in Geography, DOI 10.1007/978-3-319-28457-6

Hofmann-Wellenhof B (1997) Götterdämmerung in der Geodäsie: Verlieren Koordinaten ihre Unsterblichkeit? Österr. Z. f. Vermessung u. Geoinformation 3(2):95–102

Charamza F (1975) K identifikaci chybného bodu při výpočtu klíče podobnostní nebo afinní transformace v rovině. Geodetický a kartografický obzor 21/63(10):277–282

Charvát O (1960) Vybudování Jednotné trigonometrické sítě na území Československa. Geodetický a kartografický obzor 48(3):45–51

Chrzanowski A (1996) Recommendations of the 6th FIG symposium on deformation measurements. Wissenschaftliche Arbeiten der Fachrichtung Vermessungswessen der Universität Hannover, No. 217

Illner I (1983) Freie Netze und S-transformation. Allgemeine Vermessungs-Nachrichten 90 (5):157–170

Ingeduld M (1985) Determination of position changes of geodetic network points on base of repeated measurements. Technical papers, Series GK4/85, Technical University, Praha

Ingeduld M, Ratiborský J (1988) Řešení polohových změn bodů lokální geodetické sítě z opakovaných měření. Geodetický a kartografický obzor 76(6):140–146

Jakub V (2001) Posudzovanie stability geodetických bodov. Dissertation thesis, The Technical University of Košice, F BERG, Košice

Jindra D (1990) K problematice analýzy identit v polohových sítích. Geodetický a kartografický obzor 78(10):251–256

Just Ch (1979) Statistische Methoden zur Beurteilung der Qualität einer Vermessung. Mitteilungen/Institut für Geodäsie und Photogrammetrie, Eidgenössische Technische Hochschule Zürich, No. 27

Koch KR (1983) Ausreissertests und Zuverlässigkeitsmasse. Vermessungswesen und Raumordnung 45:400–411

Koch KR (1975) Ein allgemeiner Hypothesentest für Ausgleich-ungsergebnisse. Allgemeine Vermessungs-Nachrichten 52(10):339–345

Koch KR (1988) Parameter estimation and hypothesis testing in linear models. Springer, Berlin

Koch KR (1985) Test von Ausreissern in Beobachtungspaaren. Zeitschrift fur Vermessungswesen 110(1):34–38

Koch KR (1975) Wahrscheinlichkeitsverteilungen für statistische Beurteilungen von Ausgleichungsergebnissen. Mitteilungen aus dem Institut für Theoretische Geodäsie der Universität Bonn, No. 38, Bonn

Kok J (1984) On data snooping and multiple outlier testing. Technical Report, U.S. Department of Commerce, National Oceanic and Atmospheric Administration, National Ocean Service, Charting and Geodetic Services, Rockville, MD

Kubáček L (1996) Confidence regions in Helmert transformation. Studia Geophysica et Geodaetica 10(2):124–136

Kubáček L (1970) Some statistical aspects of the estimation of parameters of a linear conform transformation. Aplikace matematiky 15(3):190–206

Kubáček L et al (1987) Probability and statistics in geodesy and geophysics. Elsevier Science Ltd., Amsterdam

Kubáčková L (1984) Test na overenie lineárnej hypotézy o parametroch v lineárnom regresnom modeli. Geodetický a kartografický obzor 72(8):205–211

Labant S, Weiss G, Kukučka P (2011) Robust adjustment of a geodetic network measured by satellite technology in the Dargovských Hrdinov suburb. Acta Montanistica Slovaca 16 (3):229–237. ISSN 1335-1788

Lachapelle G et al (1982) Least-squares prediction of horizontal coordinate distortions in Canada. Bull géodésique 56(3):242–257

Lehmann R (2013). On the formulation of the alternative hypothesis for geodetic outlier detection. J Geodesy 87(4):373–386. ISSN 0949-7714

Lenzmann L (1984) Zur Aufdeckung von Ausreissern bei überbestimmter Koordinatentransformationen. Zeitschrift fur Vermessungswesen 109(9):474–479

Mierlo van J (1980) Free network adjustment and S-transformation. Veröff. d. Deutschen Geodät. Kommission, Reihe B, No. 252, München

Michalčák O et al (1978) Príprava výstavby diaľnično-železničného mosta cez Dunaj v Bratislave. Geodetický a kartografický obzor 66(7):161–165

Niemeier W (1980) Kongruenzprüfung in geodätischen Netzen. In: Pelzer H (Hrsg.) Geodät. Netze in Landes-und Ingenieurvermessung, Wittwer, Stuttgart

Niemeier W (1979) Zur Kongruenz mehrfach beobachteter geodätischer Netze. Fachrichtung Vermessungswesen: Wissenschaftliche Arbeiten der Fachrichtung Vermessungswesen der Universität Hannover, vol 88, Universität Hannover

Pelzer H (1980) Statistische Testverfahren. In: Pelzer H (Hrsg.) Geodät. Netze in Landes-und Ingenieurvermessung, Wittwer, Stuttgart

Pelzer H (1971) Zur Analyse geodätischer Deformationsmessungen. Deutsche Geodätische Kommission bei der Bayerischen Akademie der Wissenschaften, Reihe C, vol 164, München

Polak M (1984) Examination of reference points in distance- and combined angle-distance networks. In: Hallermann L (Hrsg.): Beitr. z. II. International symposium on deformations-messungen, Bonn, pp 426–435

Pope AJ (1976) The statistics of residuals and the detection of outliers. NOAA Technical Report NOS 65 NGS 1, U.S. Dept. of Commerce, National Oceanic and Atmospheric Administration, National Ocean Survey, Geodetic Research and Development Laboratory, Rockville, MD

Radouch V (1983) Některé aspekty testování nulových hypotéz v geodézii. Geodetický a kartografický obzor 71(4):83–87

Rangelova E, Fotopoulos G, Sideris MG (2009) On the use of iterative re-weighting least-squares and outlier detection for empirically modelling rates of vertical displacement. J Geodesy 83 (6):523–535. ISSN 0949-7714

Riečan B et al (1983) Pravdepodobnosť a matematická štatistika. Alfa, Bratislava

Sabová J, Jakub V (1999) Stability investigation of geodetic points ("Analýza polohovej stability geodetických bodov"). Acta Montanistica Slovaca 4(4):325–327

Sabová J, Jakub V (2007) Geodetic deformation monitoring, 1st edn. Košice: editor centre and editorial office AMS, F BERG, Technical University of Košice, 128 p. ISBN 978-80-8073-788-7

Schuh WD (1987) Punkttransformationen unter Berücksichtigung lokaler Klaffungsverhältnisse. Österr.Zeitsch. f. Verm. und Photogrammetrie 75(3):104–121

Skořepa Z, Kubín T (2001) Znovu o testovaní hypotéz a spolehlivosti v geodézii. Geodetický a kartografický obzor 89(3):89–94

Stichler S (1982) Untersuchungen von Massnahmen zur Beseitigung von ``Verschmierungseffekte' bei der Lagefestpunktanalyse in kleinen Strecken-und Winkel-Strecken Netzen. Vermessungstechnik 30(1):1–10

Stichler S (1985) Untersuchung von Methoden der Identifizierung stabiler Punkte als Bestandteil der geodätischen Deformationsanalyse. Vermessungstechnik 33(11):203–205

Stichler S (1981) Zur Lagefestpunktidentifizierung in kleinen Strecken- und Winkel-Strecken-Netzen mittels Verfahren nach Niemeier und Polak. Vermessungstechnik 29(11):374–376

Sütti J (1999) Quality of the connected networks with a view to their point homogeneity ("Kvalita polohovej siete z hľadiska homogenity bodov"). Acta Montanistica Slovaca 4(1):49–56

Sütti J (1996) O kompatibilite geodetických sietí. Zborník: Aktuálne problémy inžinierskej geodézie, STU Bratislava, pp 24–31

Sütti J et al (1997) Interpretácia a riešenie súradnicových rozporov pri transformáciách. Práce Katedry geodézie a geofyziky TU, No. 3, The Technical University of Košice, Košice

Sütti J et al (2000) Investigation of horizontal identity for geodetic points ("Skúmanie polohovej identity geodetických bodov"). Acta Montanistica Slovaca 5(2):121–129

Štroner M, Třasák P (2014) Outlier detection efficiency in the high precision geodetic network adjustment. Acta Geodaetica et Geophysica 49(2):161–175. ISSN 2213-5812

Teunissen PJ (1986) Adjustment and testing with the models of the affine and similarity transformations. Manuscripta Geodaetica 11(3):214–225

Vykutil J (1982) Vyšší geodézie, 1st edn. Kartografie, Praha

Weiss G, Labant S, Weiss E, Mixtaj L, Schwarczová H (2009) Establishment of local geodetic nets. Acta Montanistica Slovaca 14(4):306–313. ISSN 1335-1788

Weiss G, Jakub V (2007) The test verification of 3D geodetic points and their changes. Acta Montanistica Slovaca 12(3):612–616. ISSN 1335-1788

Welsch WM (1983) Einige Erweiterungen der Deformations ermittlung in geodätischen Netzen durch Methoden der Strainanalyse. In: Joó J, Detreköi Á (eds) Deformation measurements. Akadémiai kiadó, Budapest, pp 83–97

Welsch WM, Heunecke O, Kuhlmann H (2000) Auswertung geodätischer Überwachungs-messungen. Wichmann, Heidelberg

Werner H (1983) Beitrag zur Identifizierung von Festpunkt-bewegungen. In: Joó I et al (eds) Deformation measurements. Akadémiai kiadó, Budapest, pp 99–105

Printed in the United States
By Bookmasters